Quality Assurance and the Law

Quality Assurance and the Law

Richard Reeves and Elaine Pritchard

BUTTERWORTH
HEINEMANN

OXFORD   AUCKLAND   BOSTON   JOHANNESBURG   MELBOURNE   NEW DELHI

Butterworth-Heinemann
Linacre House, Jordan Hill, Oxford OX2 8DP
225 Wildwood Avenue, Woburn, MA 01801-2041
A division of Reed Educational and Professional Publishing Ltd

&R A member of the Reed Elsevier plc group

First published 1999

British Library Cataloguing in Publication Data
A catalogue record for this book is available from the British Library

Library of Congress Cataloguing in Publication Data
A catalogue record for this book is available from the Library of Congress

ISBN 0 7506 4176 2

Composition by Genesis Typesetting, Rochester, Kent
Printed and bound in Great Britain by
Biddles Ltd, Guildford and King's Lynn

PLANT A TREE

BTCV
British Trust for
Conservation Volunteers

FOR EVERY TITLE THAT WE PUBLISH, BUTTERWORTH-HEINEMANN
WILL PAY FOR BTCV TO PLANT AND CARE FOR A TREE.

Contents

Preface

Over the last twenty-five years there has been a considerable increase in the awareness of quality-related issues. In the world of business and commerce, this awareness has manifested itself in the in the development of what was the British Quality Standard BS 5750 into what is now the international standard BS EN ISO 9000. Along side of this, consumers in general have developed increasingly demanding expectations with regard to the quality of goods and services available in the market place. During a similar period there has also been an increase in legislation, together with an expansion of the common law, which has strengthened the protection already afforded to the consumer.

In writing this book the authors had two main aims: firstly, to provide quality practitioners, managers and those with a general interest in quality, with an insight into the legal issues involved. Secondly, to show how the implementation of a Quality Assurance Management System – such as that required in order to be registered as *a firm of assessed capability*, in accordance with BS EN ISO 9000 – can act as an aid to those businesses seeking to comply with their legal obligations.

In addition, for those following a formal course of study, the contents should prove to be particularly useful to students undertaking the Institute of Quality Assurance's Associate Membership examination, Principles and Techniques of Quality Assurance. It should also prove useful to BTEC students.

In compiling this book the authors have stated the law as they believe it to be as of November 1998.

Elaine Pritchard
Richard Reeves

vii

Acknowledgements

Extracts from BS EN ISO 9001: 1994 are reproduced with the permission of the BSI under license no. PD\1998 1423. Complete editions of the standard can be obtained by post from BSI Customer Services, 389 Chiswick High Road, London W4 4AL.

The authors would particularly like to thank Michael Forster and Rebecca Hammersley of Butterworth-Heinemann for their help in the publication of this book.

1

Sources and Classifications of English Law

Whilst there are a number of sources or origins from which English law is derived, this book is concerned with only three. The first and most commonly known is legislation which is created by Act of Parliament. The second is case law or judge made law, and the third source is the European Community.

As one or more of these sources are concerned with legislation discussed in this book, those involved with quality management should at least have some understanding of the principles and mechanisms involved.

Legislation

In the United Kingdom, Parliament is the ultimate source of law. Once an Act of Parliament, known as a statute, has been passed and receives its Royal Assent – the approval of the Monarch – it becomes the law of the land. From that point onwards, that law must be enforced in the courts, irrespective of any contrary binding precedents which may be in existence. As a result of the 'Sovereignty of Parliament', law which results as an Act of Parliament cannot be challenged on the grounds of its own legality and therefore stays in force until such time as it may be amended or repealed. However, whilst statutes, once passed, cannot be challenged on the grounds of legality, they do remain open to interpretation. Many Acts do themselves contain a section on the interpretation of clauses and phrases which appear within the Act. In addition, the Interpretation Act of 1978 defines terms and phrases which commonly occur in Acts of Parliament. In court cases where the meaning of an Act appears to be obscure or open to question, it befalls the courts to provide the necessary interpretation. In so doing the court sets a legal precedent which, with certain exceptions, is bound to be followed by other courts (see Table 1.1).

Parliament also produces what is known as delegated or subordinate legislation. This type of statute confers on Ministers of the Crown, and Government departments, the power to make or amend laws which then become binding upon society and the courts. By creating what is called an enabling Act, Parliament gives authority to Ministers or subordinate departments, which in turn allows that department to create laws, or rules, to cover specific situations with which the Act is concerned. An example of this is the Health and Safety at Work etc. Act. 1974.

Case Law or Judge Made Law

These terms cover the setting of judicial precedent. During the course of a particular trial or court case, a judge may make a decision with regard to a particular aspect of the law, its interpretation, or its application in respect of the particular circumstances of the case being heard. In delivering their judgment (making the decision) judges will:

1 state the facts of the case;
2 discuss the law concerning the legal issues raised in the case;
3 discuss the law relating to the particular facts;
4 give their reasons for reaching a particular decision.

Once this is done, the point of law upon which the case was based becomes the legal precedent, which in turn binds other judges when dealing with a similar case, or with cases concerning that particular aspect of the law. An example of this is highlighted by the case of *Donaghue* v. *Stevenson* (1932).

The facts of the case were: the plaintiff, a woman by the name of Donaghue, was given a glass of ginger beer which had been poured from a bottle. As the bottle itself had been made from dark coloured glass it was not possible to see the contents clearly. Upon replenishing her glass the defendant discovered the bottle contained the decomposing body of a snail. As a result of drinking the ginger beer the plaintiff developed gastric illness as well as suffering from nervous shock. She subsequently sued the manufacturer – Stevenson – for compensation on the grounds that her resulting condition was the result of negligence.

After initially being rejected in the courts, the House of Lords ruled that the manufacturer was liable to Donaghue in the *tort of negligence* (see Chapter 3). In delivering judgment, Lord Atkin defined who it was that was actually owed a duty of care, when he said: 'You must take reasonable care to avoid acts or omissions which you can reasonably foresee would be likely to injure your neighbour.' To establish the definition of the term neighbour, Lord Atkin went on to say: '. . . persons who are so closely and directly affected by my act that I ought to have them in reasonable contemplation as being affected when I am directing my mind to the acts or omissions which are called into question.'

In delivering the judgment, Lord Atkin not only defined to whom a duty of care was owed, he also gave rise to what is now referred to as the reasonable man test. The case set a legal precedent which is considered to be one of the most important of the twentieth century. As a result, *Donaghue* v. *Stevenson* has subsequently become the leading and most often quoted case concerning negligence in English law.

Whilst legal precedent becomes binding on judges and courts dealing with subsequent cases, there are however a number of exceptions, the most notable of which being those which are dealt with by the House of Lords. As the highest appeal court in the United Kingdom, concerned with both criminal and civil matters, the decisions made there are binding on all the other courts. However, the House of Lords is not bound by the decisions or precedents set by any other court, nor is it bound by its own previous rulings. Within the hierarchy of the English court system there is a general rule that the lower courts are bound by the decisions of the courts above them. However, like the House of Lords, there are other courts in the English legal system which are not bound by their own previously set decisions. A more detailed explanation of the courts which are, and those which are not, bound by precedent is given in Table 1.1.

European Community Law

The European Communities Act of 1972 created a new source of English law. As a result of the Act all courts in the United Kingdom have to recognize EC law, irrespective of whether it is a direct result of treaties between the member states, or produced by EC legislation.

Within the structure of the EC the main institutions are:

1 the European Parliament, which consists of varying numbers of representatives – MEPs – from each member state;
2 the Council of Ministers, which has a single representative from each state;

Table 1.1   Courts having precedent

Name of court	Whether or not bound by its own previous decisions.	Bound by the decisions made by:
House of Lords	No	Not applicable
Court of Appeal (criminal)	Not in all cases	House of Lords
Court of Appeal (civil)	Yes	House of Lords
Queen's Bench Division of the High Court:		
– (criminal)	Not in all cases	House of Lords Court of Appeal (in some circumstances)
– (civil)	Yes	House of Lords Court of Appeal
Crown Court	Varies with circumstance	House of Lords Court of Appeal Divisional Courts of the High Court
The Inferior Courts: – Magistrates (criminal) – County Court (civil)	These courts are not bound by their own decisions nor may they set legal precedent	All the superior courts

3 the European Commission, which also has a single representative from each state.

Of the three, it is the Commission alone which is tasked with the role of instigating and proposing legislation which, once enacted by the European Parliament, is subsequently presented to the Communities' member states in the form of either a regulation, a directive or a decision.

Regulations

Regulations are automatically binding upon all the member states. In the United Kingdom, EC regulations are binding, at least in theory, without prior reference being made to Parliament. However, there will invariably be some Parliamentary discussion on the topic, or possibly the need for legislative action to be taken. This may well require the repealing or amendment of current English law which itself would be contrary to the regulation concerned.

Directives

As with regulations, EC directives are binding upon all member states. However, each individual state may choose to implement the directive by whatever means they choose, providing that this is accomplished within the specified time. An example of this is the Consumer Protection Act 1987, which itself implements an EC directive of 1985 concerning Product Liability.

Decisions

Decisions are directed at individual member states or corporations, and as a result they become binding upon those states or corporations. Once made, the member state concerned is required to implement the decision immediately, although to do so may require some legislative action.

The European Court

The Court of Justice of the European Communities – the European Court – is concerned with the implementation and, where necessary, the interpretation of European treaties. In these matters its decision is final and binding upon all of the member states, which must in turn enforce the decision through their own national courts. Under English law any court may request the European court to give a ruling, however not all of them are bound to do so. Indeed both the High Court and the Court of Appeal have the right to interpret Community law, and neither are bound to grant the right of appeal to the European Court. The House of Lords is the only English court which must grant the right of appeal to the European Court if one of the parties involved in the case requests it.

Civil and Criminal Law

The law is classified as either private or public law, but is more commonly known by the terms civil and criminal law.

Civil Law

Civil law is, at first glance, the area with which the quality practitioner would be predominantly concerned, dealing as it does with the rights of individuals, and companies, in such areas as contract and negligence. Civil law provides a system whereby an individual or company can seek redress or claim compensation for injury or loss that has been suffered.

Heard in the civil courts, the proceedings are instigated privately and referred to by the names of the parties involved, e.g. *Donaghue* v. *Stevenson* (1932) refers to Donaghue the plaintiff – the person making the complaint, and Stevenson the defendant – the person against whom the complaint is being made. The letter 'v.' is the abbreviation for 'versus', a term which is now commonly replaced by the word 'against'. The year (1932) refers to the year in which the case appears in the law reports.

Criminal Law

A criminal offence, whilst perpetrated against an individual or business, is taken so seriously as to be considered a crime against the state. As a result, criminal proceedings, which are serious enough to be tried on indictment – in the Crown Court – are instigated by the state, and prosecuted in the name of the monarch. Currently criminal cases are referred to as Regina versus the defendant, e.g. Regina versus Smith, which is usually written as *R* v. *Smith*. Cases which are tried summarily in the Magistrates Court may well bear the name of a private person in place of Regina. An example of this is the case of *Ellis* v. *Wings Ltd.*, where a prosecution was brought in the name of David Kenneth Ellis who was at the time the county trading standards officer for Devon. This particular case also serves to highlight what seems to be a peculiarity of the designation of case titles, for upon appeal to the Queen's Bench Division of the High Court the names of the parties involved were reversed and the case became known as *Wings* v. *Ellis*, the name which it then continued under until it reached its conclusion in the House of Lords in 1985 (see Fair Trading, Chapter 6).

Where guilt is proven, the guilty party can be punished by fines, terms of imprisonment or both. A guilty party can also be ordered to make restitution to the victim of the crime. In the event of a violent crime, where the guilty party is unable to make restitution, it may be possible for the victim to claim compensation, for any injuries they sustained, from the Criminal Injuries Compensation Board.

The most obvious of crimes, such as theft and murder, would appear to make criminal law of little professional importance to those involved in

Table 1.2 Differences between civil and criminal liabilities

	Civil liability	Criminal liability
To whom liability is owed.	Any individual or business which has suffered injury or loss through such causes as negligence	The injury of an individual caused by a defective product is considered to be so serious that it also constitutes an offence against the state. As a result the state may decide to prosecute the manufacturer or the seller. See consumer protection, Chapter 4
Penalties	The defendant can be compelled to pay damages (a sum of money) as compensation to the plaintiff	The defendant can be ordered to pay a fine – payable to the state – or be sent to prison, or both
Purpose	To compensate an individual or company for the injury or loss suffered	To punish, and deter the guilty from further offences
Where the case is heard.	In the civil courts	In the criminal courts

quality assurance. However a good deal of what appears, at first glance, to be civil legislation actually contains provision for criminal proceedings. The Consumer Protection Act 1987 (see Chapter 4), and the Food Act 1990 (Chapter 7) both contain provision for criminal proceedings which can lead to fines for companies or individuals, or terms of imprisonment for those held responsible.

For reference, the differences between civil and criminal liability, the purposes of each system, and the penalties that can be imposed, are summarized in Table 1.2.

Civil Courts

County Courts

The County Court is the first of the civil courts, with wide ranging jurisdiction with regard to settling civil issues. However, as far as the quality practitioner

Table 1.3   Financial boundaries for the jurisdiction of the County and High Courts

Type of claim	County Court	Queen's Bench Division of the High Court
Personal injury	Cases where the claim for damages is less than £50,000	Cases where the claim for damages is £50,000 or more
Other than personal injury	Claims for damages of less than £25,000	Claims for damages of £50,000 or more

is concerned, it will be the court's jurisdiction with regard to areas such as: damages for negligence, product liability and breach of contract that will be of interest. The jurisdiction of the courts in cases where personal injury is concerned, is defined by the amount of damages which are being claimed, although there are areas of overlap between the County Court and the High Court. The financial boundaries for the jurisdiction of the County and High Courts are shown in Table 1.3.

It should be noted that claims for damages arising from other than personal injury can be heard either in the County or the High Court. However there are a number of factors which would necessitate such a case being heard in the High Court :

1 Where the issues raised in the case were considered to raise matters of general public importance.
2 In cases where the facts, and the legal issues involved are extremely complex.

An additional consideration would be whether or not the County Court would be able to give a hearing to the case sooner than the High Court. However this cannot be considered the prime reason for transferral of a case to the County Court.

This list is far from complete, but it does contain those areas most likely to be of professional concern to the quality practitioner.

As can be seen from Figure 1.1, the High Court is above the County Court in terms of the legal hierarchy. However, unlike criminal courts where cases are passed up, civil cases may be passed down from the High Court to the County Court; as well as up from the County Court to the High Court. In all cases, appeals from the County Court always go directly to the Court of Appeal (Civil Division).

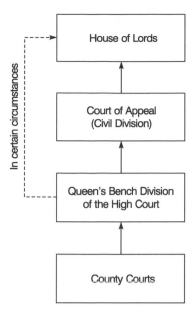

Figure 1.1 Courts with civil jurisdiction

It is worth noting that the term 'small claims court' refers to an area of work which is actually part of the County Court's jurisdiction. Presided over by a judge, without a jury, the court aims to be less formal, with the parties involved being encouraged to present their own case without the aid of a legal representative such as a solicitor. To facilitate this arrangement the strict rules of evidence are dispensed with. The role of the judge becomes more that of an arbiter, although they do exercise the necessary control to ensure that all the parties involved receive a fair hearing. The maximum claim which can be made in this way is currently £3000.

The High Court

The Queen's Bench together with the Chancery and Family divisions form the three divisions of the High Court, which has both civil and criminal jurisdiction. The High Court has unlimited jurisdiction in civil cases, and although each division specializes in their own specific areas, they are all empowered to hear any action brought in the High Court. The majority of civil cases heard by the Queen's Bench deal with either contracts (see Chapter 2) or tort, e.g. negligence (Chapter 3).

The Court of Appeal (Civil Division)

As its name suggests the Court of Appeal hears appeals against decisions made in the High Court or the County Courts. In its civil capacity the majority of appeals are made on points of law, although appeals may be heard which dispute the amount of damages or costs which have been previously awarded. The Court of Appeal has the power to:

1 allow or dismiss the appeal;
2 reassess the amount of damages or costs which have been previously awarded;
3 order a new trial.

The House of Lords

The House of Lords is the final court of appeal in the United Kingdom, although provision is made for law lords from Northern Ireland or Scotland to be present for cases which originate from their respective areas. In all cases, irrespective of which court grants the right of appeal, that court must certify that the case concerned involves a point of law which is of general importance to the public. Whilst appeals to the House of Lords are generally granted by the Court of Appeal, there is provision for appeals which originate in the High Court to by-pass or 'leapfrog' the Court of Appeal. An example of where leapfrogging is most likely to occur would be cases which involve patents, or revenue appeals. This type of case containing issues where the construction of the statute is involved is usually extremely complex.

Criminal Courts

Irrespective of the nature or the severity of the alleged offence, all criminal proceedings receive their first hearing in the Magistrates Court. This court, which does not have a jury, is presided over by between two and seven lay magistrates, with legal advice being provided by the clerk to the court.

Alternatively a full time stipendiary magistrate, who is a solicitor or barrister, may preside alone. Magistrates have the authority to try a number of offences, including those charged under the Consumer Protection Act 1987, the Food Act 1990, and the Health and Safety at Work Act 1979, each of which provides provision for criminal prosecution.

A Magistrates Court has a number of options when dealing with criminal offences.

1 Offences of a relatively minor nature can be tried, decided and sentence imposed. In such cases the maximum penalty is six months imprisonment for a custodial sentence, or a fine of £5000 for any single offence. However fines of up to £20,000 can be imposed for offences concerning health and safety.
2 For an indictable offence, one that has to be tried by jury, the magistrate must pass the case on to the Crown Court.
3 In some instances a Magistrates Court may try and convict, and then send the guilty party to the Crown Court for sentencing, the Crown Court being able to decide upon more severe sentences than a Magistrates Court.

For those convicted by a Magistrates Court, two possible routes of appeal exist (see Figure 1.2).

1 The Crown Court can hear appeals against conviction – providing that the accused has not pleaded guilty – or against the severity of the sentence. However, in the case of the latter it should be borne in mind that the Crown Court could also impose a more severe sentence.

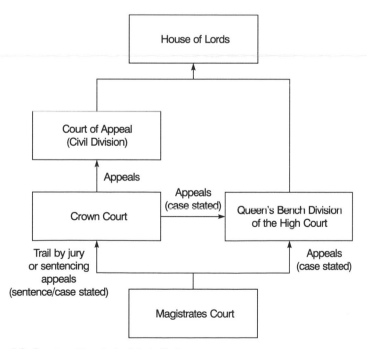

Figure 1.2 Courts with criminal jurisdiction

2 The Queen's Bench Division of the High Court may consider an appeal by means of case stated. This in essence is a written appeal where the judgment of the magistrate is challenged on the grounds that it is wrong in law.

The Crown Court

The Crown Court is the successor to the Courts of Assizes and the Quarter Sessions. It has jurisdiction to try all indictable offences. It also hears appeals and transfers for sentence from the Magistrates Court.

The Queen's Bench Division of the High Court

This court conducts appeals on points of law – case stated – from both the Magistrates and the Crown Court. In addition the High Court has a supervisory role over inferior (Magistrates) courts, which permits it to discipline those courts, and put right their mistakes.

The Court of Appeal (Criminal Division)

The criminal division hears appeals made by persons convicted and sentenced by the Crown Court, or convicted by a magistrate and subsequently sentenced by the Crown Court. The Court of Appeal has the power to:

1 overturn (quash) a conviction;
2 reduce or increase sentence;
3 order a new trial to be conducted.

The Court of Appeal will also consider cases which have been referred to it by the Attorney General, when he considers the sentence to be too lenient.

The House of Lords

As with civil cases, the House of Lords is the final court of appeal in the United Kingdom. In considering appeals from the Crown Court or the High Court it uses the same criteria as for civil cases, viz. these courts must confirm that the case involves a point of law of general importance to the public.

2

Contract Law

Contracts are agreements made between two or more parties, which are legally enforceable. To ensure that contracts reflect the intention of all parties concerned, it is important that the contract is drawn up very carefully and that all parties are satisfied with the contents. A breach of contract by one of the parties can be expensive if the party who has breached the contract is required to pay damages to the other party to the contract. *Following a quality procedure from the inception of the contract can help to prevent any misunderstandings later on and can help to promote good business relationships between contracting parties. For the fundamental requirements of a quality management assurance system see contract review 4.3.2 of BS EN ISO 9001.*

It is important to understand the legal implications of entering into a contractual relationship with another party, so that each party's obligations under that contract can be fully appreciated.

What is an Agreement?

It has already been stated that a contract is an agreement between two or more parties, but just what is meant by *agreement*? Before a court can decide if there has been a breach of contract, it needs to be ascertained:

1 that there is indeed a contract in existence;
2 when the contract came into existence;

3 what the terms of the contract are; and
4 which party failed to meet their contractual obligations.

To help the court in this process, the making of the contract is broken down into different components. This is essential because contracts do not have to be written down and signed by both parties before they are legally enforceable. It is possible to have verbal contracts between parties (except in cases such as contracts for the sale of land) which are enforceable by the courts. Contracts which are written down may not all be contained in one document. The contract may include: several pieces of paper, letters, faxes, telephone conversations and maybe even conversations at meetings. It can be seen, therefore, that it is important to understand how a contract is formed in order to ensure that each agreement entered into is meant to be legally enforceable. Also, to ensure that 'pre-contract negotiations', remain just 'negotiations' without becoming part of the contract; or are not actionable misrepresentations.

The following, are the essential features of a valid contract:

■ offer
■ acceptance
■ consideration
■ intention
■ legality
■ capacity

Each of these will now be considered in turn.

Offer

To enable a contractual agreement to be reached, it is necessary for the parties to the contract to negotiate the terms on which the contract will be based. The terms may include the specification, the price and the delivery date. During these discussions it may not be intended that any statements made will form part of the contract, or that the parties will be legally bound by their statements. But, as will be seen later in this chapter, there are circumstances when this will indeed be the case. An agreement between the parties will be reached, when one party makes an 'offer' to the other party, who then 'accepts' that offer. An offer is a statement of terms, proposed by the offeror to the offeree, which the offeror intends to become legally binding as soon as they are accepted by the offeree. The terms must be unequivocal, definite terms to which the offeree can readily assent.

Distinguish between an 'offer' and an 'invitation to treat'

However, not all statements made by the parties are either 'offers' or 'acceptances'. Many statements are made during negotiations, to encourage the other party to the contract to make an offer. These statements are known as 'invitations to treat', and are not legally binding on either party. An example of this can be found in the case of:

Harvey v. *Facey* (1893)

The plaintiffs telegraphed the defendants, saying, 'Will you sell us Bumper Hall Pen? Telegraph lowest cash price.' The defendants telegraphed in reply, 'Lowest price for Bumper Hall Pen, £900' The plaintiffs then telegraphed, 'We agree to buy Bumper Hall Pen for £900 asked by you. Please send us your title-deeds.' There was no other communication. The plaintiffs then brought an action against the defendants for breach of contract. The court held that there was no breach of contract. The reason for the court's decision was that the first two telegraphs were merely negotiations (invitations to treat). The defendant's reply to the plaintiff's question was not an *offer* to sell Bumper Hall Pen, as claimed by the plaintiffs. The defendants had merely indicated the lowest price they were prepared to accept *if* they decided to sell. Therefore, the plaintiff's next telegraph could not be an *acceptance*, as there was no offer to accept. No binding contract had been made by the parties as there was no offer and corresponding acceptance. Therefore, there could be no action for breach of contract.

It can be seen that there may be difficulty in establishing whether there is an 'invitation to treat', or an 'offer' in existence. Over the years, the courts have attempted to distinguish between situations resulting in an 'invitation to treat', and those resulting in 'offers'. The following precedents have been set.

Advertisements

The general rule regarding advertisements is that they are invitations to treat and not offers. In the case of:

Partridge v. *Crittenden* (1968)

The appellant had placed an advertisement in a magazine which stated, 'Bramblefinch cocks and hens, 25s each'. Under legislation for the protection of wild birds it was unlawful to *offer for sale* wild birds. The appellant was charged under this legislation. The court found the appellant was not liable for this statutory offence as the advertisement was an invitation to treat and not an *offer to sell*.

Similarly, in the case of:

Harris v. *Nickerson* (1873)

The defendant auctioneer had advertised an auction which was to be held in the near future which contained lots of certain furniture. The plaintiff attended the auction with the intention of buying some of the furniture advertised. The furniture was not included in the auction and the plaintiff sued the auctioneer for damages, to include the cost of travelling to the auction and for the loss of his time (taken to get to the auction). His action failed as the advertisement was held to be merely an invitation to treat and not an offer to sell the furniture. His claim would only have been successful if there had been a legally binding contract between the parties.

Having just stated that advertisements are generally invitations to treat and not offers to sell, there may, of course, be exceptions to this general rule. This is illustrated by the case of:

Carlill v. *Carbolic Smoke Ball Company* (1893)

The Carbolic Smoke Ball Company placed an advertisement in a newspaper for a product known as a 'smoke ball'. The advertisement further stated that £100 would be paid to anyone who bought a smoke ball and used it in accordance with the instructions and still contracted influenza. The company also stated that, in order to show its 'sincerity in the matter', £1000 had been deposited in the bank. The plaintiff, after seeing the advertisement in the newspaper, bought a smoke ball and used it in accordance with the instructions. Despite this, the plaintiff contracted influenza, and sued the company for £100.

If the courts had followed the general rule, that advertisements were invitations to treat, then Mrs Carlill would not have received the £100 she was claiming, as there was no contractual obligation on behalf of the company to pay her. The court, however, decided that this particular type of advertisement was in fact an offer. The information was clear and complete enough to constitute an offer; the intention to enter into legal relations was shown by the money placed in the bank, 'as a sign of sincerity'; the plaintiff had accepted the offer by buying the smoke ball, which was also the consideration for the defendants promise to pay £100 if the smoke ball did not prevent the plaintiff from catching influenza. As all the essential elements required for contract formation were present, then a contract was in existence, and the company was legally obliged to pay Mrs Carlill £100.

Advertisements of this kind give rise to 'unilateral' contracts. Most contracts entered into are 'bilateral contracts'.

An example of a bilateral contract is as follows. Person A offers to sell goods to person B for a certain price. Person B agrees to buy the goods from person A for the agreed price. Both parties incur obligations under this contract. Person A has an obligation to sell the goods, and person B to buy the goods. If either party refused to carry out their part of the contractual agreement, the other party would be entitled to sue for breach of contract.

In a unilateral contract, only one party incurs legal obligations. Taking the case of Carlill as an example, the company incurs liability by promising to pay £100 to persons buying and using the advertised product, as instructed, who then contracted influenza. Persons reading the advertisement were under no obligation whatsoever. They may buy the product if they wish. The choice is theirs. The company incurs no liability at the outset, by merely placing the advertisement in a newspaper, but it must be realized that as soon as someone buys the product, and proceeds to use it as instructed, then the possibility of liability arising exists.

Reward cases are other examples of unilateral contracts. If a bank placed an advertisement in a newspaper, offering a reward for information supplied, which led to the apprehension and subsequent arrest of persons responsible for robbing the bank, then as soon as the conditions in the reward offer had been complied with, the bank would be bound to pay the reward money to those providing the information. Notice, later on, the differences to the rules on acceptance of an offer and the requirements of consideration when dealing with unilateral contract as opposed to bilateral contracts.

From the above discussion, it can be seen how important it can be when placing an advertisement that words are chosen carefully. Regulation of advertisements is carried out by the Advertising Standards Authority, whose aim is to ensure advertisements are truthful and not offensive, but, as shown above, liability may also be incurred under contract law resulting in a breach of contract action being taken against those unwilling to fulfil their obligations.

Auctions

In the case of *Payne* v. *Cave* (1789), which is discussed in greater detail later, it was established that the goods for sale are invitations to treat. The bids made are offers to buy those goods, and the acceptance is when the auctioneer brings down the hammer.

Goods on display in shops

It was established in the following case that the goods on display in shops are invitations to treat and are not goods which are being offered for sale.

Pharmaceutical Society of Great Britain v. *Boots Cash Chemists* (1953)

Under s18(1) Pharmacy and Poisons Act 1933 it is stated that 'it shall not be lawful . . . a) for a person to sell any (listed) poison, unless . . . the sale is effected by, or under the supervision of a registered pharmacist'. A Boots Chemist shop had been converted into a self-service store. A person had placed a certain bottle of poison into her basket, along with other items and paid for them all at the check-out. Boots Chemist were prosecuted under the aforementioned Act for failing to sell poison under the supervision of the registered pharmacist.

The action failed, as the court held that the goods on display in the shop were merely invitations to treat. When the shopper placed items into the basket provided, an offer was being made to buy those goods. The offer could be withdrawn at any time before acceptance by placing the goods back on the shelf. Acceptance was made by the person at the check-out accepting the money for the goods. Therefore the sale had been concluded under the supervision of the pharmacist and Boots Chemist were not in breach of the Act.

It had been argued that the display of goods in the shop was an offer to sell those goods. Placing the goods in the basket, it was claimed, was the acceptance to buy those goods. If this had been found to be the case, it would be a breach of contract for any shopper to change their minds and replace goods back on the shelf!

A practical application of this can be seen by the following example: A pair of jeans are advertised in a shop window for £4. A person enters the shop and takes a pair of the advertised jeans to the check-out. The goods appear on the till at £40. The customer claims that the shop are in breach of contract for failing to sell the jeans at £4. By applying the rules discussed previously, it can be seen that the jeans are invitations to treat, the customer has made an offer to buy the jeans for the sum of £4, which has been rejected by the cashier. As there is no corresponding offer and acceptance, there is no legally binding contract, and therefore no obligation to sell the jeans for £4.

This does not mean that it is perfectly acceptable to place incorrect prices on goods displayed in shops. The Consumer Protection Act 1987 makes it an offence to give misleading price indications. S20(1) CPA 1987 states:

> subject to the following provisions of this Part, a person shall be guilty of an offence if, in the course of any business of his, he gives (by any

means whatever) to any consumers an indication which is misleading as to the price at which any goods, services, accommodation or facilities are available (whether generally or from particular persons)

This means that incorrectly pricing goods is a criminal offence and could lead to a fine being imposed, on conviction of the offence.

Catalogues, circulars and price lists

These are all considered to be invitations to treat. The publishing of descriptions of goods, together with a price list, is merely an invitation to the public to make an offer to buy the goods. Any other interpretation would be unworkable as explained by Lord Herschell in the case of:

Grainger and Son v. *Gough* (1896)

This case concerned a price list issued by a wine merchant who was then unable to supply some of the wine advertised. Lord Herschell stated:

> The transmission of such a price-list does not amount to an offer to supply unlimited quantity of the wine described at the price named, so that as soon as an order is given there is a binding contract to supply that quantity. If it were so, the merchant might find himself involved in any number of contractual obligations to supply wine of a particular description which he would be quite unable to carry out, his stock of wine of that description being necessarily limited.

The above instances are examples of situations where it was necessary for the courts to distinguish between 'invitations to treat' and 'offers'.

Revocation of the offer

An offer can be withdrawn by the offeror any time prior to the acceptance being received. It is said in these instances that the offer has been revoked. The case which established this principle, is that of:

Payne v. *Cave* (1789)

At an auction, a bid was made and then withdrawn before the auctioneer's hammer fell. A dispute arose as to whether there was a binding contract in force. The court held that as the goods for sale at auctions were invitations to treat (they were on display inviting offers from prospective bidders), the bids were classified as offers. Persons at the auction would have viewed the goods beforehand and decided on a price they were prepared to pay. They would ensure they had all the necessary information before they started to bid, so the offer would then be firm and capable of an unconditional acceptance. The acceptance would be the bringing down of the hammer by the auctioneer. It was decided that it was perfectly legitimate for an offeror to change his mind and revoke his offer as long as this was carried out before the acceptance. In this case, the bid was withdrawn before the hammer fell, so the offer had been revoked. As there was no offer for the auctioneer to accept, there was no contract in existence.

Rewards

In the case of advertisements for a reward, there is American support for the fact that the offer contained in that advertisement may be revoked by placing a further advertisement in the same place(s) as the offer had originally been placed. This was held in the case of *Shuey* v. *US* (1875). It was further stated that it was unnecessary for all acceptees of the original offer to have read the revocation of the offer in order for it to take effect, as long as the revocation had been published in the same place as the offer.

Acceptance

An acceptance to the offer must be absolute and unequivocal. There are two aspects to acceptance: the 'fact of acceptance' and the 'communication of acceptance'.

The fact of acceptance

Acceptance may be: spoken – either face to face, or on the telephone; written in a letter, facsimile, telex or contained in any other documents which have passed between the parties. It is important to be able to establish the precise moment of acceptance of the offer in order to ascertain the exact moment the contract came into existence. There is also the possibility that acceptance may be inferred from the conduct of the offeree. This was the decision of the court in the case of:

Brogden v. *Metropolitan Railway Co.* (1877)

Brogden was a supplier of coal, and had been supplying the defendant Railway Co with coal for several years on an informal basis. The parties then decided to formalize their agreement. A draft agreement was drawn up by the Company's agent and sent to Brogden for approval. A blank space had been left in the agreement for Brogden to insert the name of an arbitrator. The name of an arbitrator was included in the agreement which was signed by Brogden, marked 'approved', and returned to the agent. Upon receipt of the agreement, the agent placed the document in his drawer, but took no further steps to complete its execution. Brogden and the Metropolitan Railway Company continued trading, but on the revised terms of the new agreement. A dispute arose between the parties, and Brogden denied the existence of a legally binding contract.

In order for there to be a legally binding contract between the parties, there had to be a mutual assent to all the terms of the contract. The difficulty for the court was to ascertain when the parties assented to the terms of their agreement. Brogden could not have been accepting the company's offer when returning the draft agreement, because he had inserted the name of an arbitrator (a term of the contract), which the company had not had the opportunity to either agree to, or reject. Alternatively, if it was argued that Brogden's returning of the draft agreement (including the name of an arbitrator) was an offer to supply coal to the Railway Company, there would then need to be an acceptance of those terms by the Company in order for there to be a contract in existence. As there was no response from the Company, there was clearly no acceptance. The court then examined the conduct of the parties and saw that both parties had behaved as though there was a contract in existence. The court concluded that the parties had both assented to the terms of the agreement and had intended to be bound by those terms. The House of Lords decided that a contract had come into existence either, when the Company had ordered its first load of coal from Brogden, or when Brogden delivered the order.

Acceptance in the case of unilateral contracts is inferred by the conduct of the acceptee. Take the case of Carlill as an example again. Mrs Carlill did not have to write and inform the Carbolic Smoke Ball Company that she was going to buy a smoke ball and use it as they had instructed. The fact that she had bought the smoke ball was sufficient to constitute an acceptance.

Acceptance or counter-offer?

If one or more terms of an offer were not accepted unconditionally by the offeree, but were amended in some way, then this would not amount to an acceptance, and no contract would exist. The amended offer is called a 'counter-offer' and it would now be up to the original offeror to accept or reject this 'counter-offer'. A good example of this situation can be seen in the case of:

Hyde v. *Wrench* (1840)

On 6 June, the defendant offered to sell an estate to the plaintiff for £1000. In reply to this offer, on 8 June, the plaintiff offered to buy the estate for £950. On 27 June the defendant decided to reject the plaintiff's offer. The plaintiff then wrote to the defendant on 29 June to accept the defendant's original offer of £1000. The defendant no longer wished to sell the estate to the plaintiff. The plaintiff brought an action against him for breach of contract. The court held that there was no contract in existence. The plaintiff's reply of £950 was a counter-offer. It was not an acceptance, because it had changed the terms of the original offer and was therefore not an unconditional acceptance to the defendant's offer.

When negotiating the terms of a contract, both parties need to read any correspondence they receive from the other contracting party very carefully in order to recognize the response as being a firm offer, an acceptance, a counter-offer, or, as in the following case, merely a request for further information.

Counter-offer or request for further information?

Stevenson v. *McLean* (1880)

On Saturday, the defendants offered to sell the plaintiffs 3800 tons of iron 'at 40s nett cash per ton, open till Monday'. The plaintiffs telegraphed the defendants early on Monday. 'Please wire whether you would accept 40s for delivery over two months, or if not, longest limit you would give.' The plaintiffs received no reply, and so, later that day, at 1.34pm the plaintiffs accepted the offer to sell at 40s cash. By this time, the defendant had already sold the iron to a third party. The plaintiffs were informed of this fact by a telegram which was sent at 1.25pm. The two telegrams crossed.

The plaintiffs sued for breach of contract. The court had to decide if a contract had come into existence, and if so, when. The offer had to have been in existence at 1.34pm when the plaintiffs sent their acceptance by telegram. If their first telegram had been a counter-offer, then the original offer would have been destroyed and there would be no breach of contract. The plaintiffs telegraph, early on Monday, could not have been an acceptance because it was not an unconditional acceptance of all the terms of the offer. The telegraph was not a counter-offer, as it did not definitely change the terms of the offer. The court decided that on the proper construction of the negotiations between the parties, was that this correspondence was merely a request for further information which the defendant should have responded to. Consequently, the original offer had remained open and was accepted by the plaintiffs by the telegram sent at 1.34pm. As acceptance by telegram takes place when placed in the control of the post office and revocation of the offer does not take place until received by the offeree it was decided that there was a binding contract in existence between the parties. **Ignore any correspondence at your peril!**

Revocation of offer

Another point which was raised in this case was that the offer made by the defendant had been revoked by the telegraph sent at 1.25pm which informed the plaintiffs that the iron had been sold to a third party. However, a revocation needs to be received by the offeree before it is effective. If the facts of this case are examined carefully, it can be seen that the telegraph revoking the offer was received by the offeree after the offeree had sent his acceptance. Therefore, a contract had already been formed and the revocation was too late.

If an offer has been made to another party and then it is decided to withdraw that offer, careful consideration must be given to the method by which the revocation is made. This important point is highlighted by the case of:

Byrne v. *Van Tienhoven* (1880)

On 1 October, the defendants (in Cardiff) wrote to the plaintiffs (in New York) offering to sell them tinplates. On 8 October the defendants wrote to the plaintiffs again, this time withdrawing their offer. On 11 October, the offer reached the plaintiffs, who immediately accepted the offer by telegram. A confirmation of the offer was also sent by letter on 15 October. The letter of withdrawal, sent by the defendants on 8 October reached the plaintiffs on 20 October. Because the withdrawal of the offer was not received by the plaintiff until after the plaintiff had accepted the offer, the withdrawal was ineffective, and a binding contract was in existence.

Posting a letter of revocation has proved to be problematic in the past (as in the above case) either because of the time the letter has taken to reach the offeree, or because the letter has been lost in the post and never arrived at its destination. Care must be taken when withdrawing any offer made. It would be much safer to ensure that being bound to the terms of the offer is in the offerors best interests before the offer is made to another party. Remember also, that any communication between contracting parties should always have a written copy which could be used in evidence if a dispute arose. Keep records of any phone calls made, the name of the other person on the phone, and the subject of the phone call. Always confirm in writing what was said. If rules such as these are incorporated into quality procedures, there is less chance of a mistake being made or a misunderstanding occurring with a contracting party.

The case below established that it did not have to be the offeror who communicated the withdrawal to the offeree, but the withdrawal may be communicated by a third party to the contract. In the case of :

Dickinson v. *Dodds* (1876)

The defendant gave the plaintiff the option to buy some land for £800, promising that this offer would stay open until 9am on 12 June. On 11 June the defendant sold the land to another party. The plaintiff heard about this from a third party, and promptly informed the defendant of his acceptance of the offer. As the land had already been sold, the plaintiff then brought an action against the defendant asking the court to order the defendant to fulfil his contractual obligation.

The decision of the court was that there was no contractual obligation between the parties because the offer to buy the land had been revoked before the offer had been accepted by the offeree and so this was an effective withdrawal of the offer. The judge decided that the plaintiff was left in no doubt that the defendant no longer wished to sell the land to him, in fact it was as clear as if the defendant had told the plaintiff himself of his decision. The plaintiff had heard from a reliable source of the defendant's decision and this was sufficient.

It is, maybe, a little unsatisfactory that an offeree may be told that the offer which has been made to him by the offeror is no longer open for him to accept if this knowledge is conveyed to him by a 'reliable source'. The interpretation of who may be a 'reliable source', may be too unreliable to be relied upon! The person making the offer, should always ensure that if he wishes to withdraw the offer, that he does so personally to avoid any doubt.

After establishing there was a 'fact of acceptance', it needs to be established that there was 'communication of acceptance'.

Communication of acceptance

The general rule in contract law is that the communication of the acceptance has taken place, only when it has been communicated to, and received by, the offeror.

Entores Ltd v. *Miles Far East Corporation* (1955)

The plaintiffs were a company in London and the defendants were an American corporation. The defendants had agents in Amsterdam. The plaintiffs made an offer to buy goods from the defendants, by telex to the agents in Amsterdam. The agents accepted this offer. The telex service consisted of teleprinters which were operated like typewriters in the one country and which were received and typed almost immediately in the other country. Later, the plaintiffs claimed that the defendants had broken their contractual agreement and wished to sue for this breach. They could only serve the writ if the agreement had been made in England. The defendants claimed that the contract had been concluded in Holland where they accepted the contract.

The court decided that the general law of contract applied, which was that, acceptance was only valid when it had been communicated to and received by the offeror. Using the telex service was equivalent to other forms of instantaneous communication and so it was effective when it was received by the offeror in London. This enabled the plaintiffs to issue the writ.

The point for the court to decide in this case was *where* the acceptance took place, but in the following case the court had to decide at what point the acceptance took place.

Brinkibon Ltd v. *Stahag Stahl und Stahlwarenhandelsgesellschaft mbH* (1983)

The plaintiffs were an English company and the defendants were an Austrian company. The defendants had offered to sell goods to the plaintiffs, who accepted the offer by telex sent to Austria. Later, when the plaintiffs issued a writ claiming damages for breach of contract, the court held that the writ should be set aside as the contract was formed in Austria, outside the jurisdiction of the court.

One of the judges stated,

> The present case is, ... itself, a simple case of instantaneous communication between principals, and, in accordance with the general rule, involves that the contract (if any) was made when and where the acceptance was received. This was on 3 May, 1979 in Vienna.

It can be seen that many of these cases were decided many years ago and technology has improved considerably since they were heard and other forms of communication are now being used. New questions are raised by the use of these new methods of communication. For example, when is acceptance effective from, if faxed through to the offeror? Is it when it appears on the machine or when it is read by someone in the office, or, when read by the person to whom it is addressed? The judge in the above case also stated the following:

> Since 1955 the use of telex communication has been greatly expanded, and there are many variants on it. The senders and recipients may not be the principals to the contemplated contract. They may be servants or agents with limited authority. The message may not reach, or be intended to reach, the designated recipient immediately; messages may be sent out of office hours, or at night, with the intention, or on the assumption, that they will be read at a later time. There may be some error or default at the recipient's end which prevents receipt at the time contemplated and believed in by the sender. The message may have been sent and/or received through machines operated by third persons. And many other variations may occur. No universal rule can cover all such cases; they must be resolved by reference to the intentions of the parties, by sound business practice and in some cases by a judgment where the risks should lie . . .

This shows, that until new cases concerning these issues come before the courts, the law will remain uncertain. Each business enterprise must ensure that careful thought is given as to what form of communication is used when entering into contractual obligations with another party. Common business practice may determine at what point the acceptance was received by the offeror. if, for example, an acceptance was sent by fax after normal office hours, then the court may decide that the acceptance had been received by the offeror the following morning. Or if an acceptance was sent out on a Friday evening it may be decided that Monday morning would be the appropriate time for the acceptance to have been received. The same principle may apply to a situation where an acceptance was sent out on the evening before the organization shut down for a holiday. The courts would have to take all the circumstances of the case into account in order to reach a decision as to the time acceptance was received. Whether, for example, it was common knowledge that the organization shut down for a holiday at that time; whether it may be common in that particular trade to shut down at a particular time, or whether the two parties had conducted business together before and it was reasonable that the offeree should know when the holiday period was.

The next problem for the court to decide would then be to determine at exactly what time the acceptance had been received by the offeror. As stated before, the law is uncertain whether acceptance would be received at maybe 9.00am. when the office opened, or maybe 10.00am. when the offeror had had chance to see the fax. Also, what may be decided if the offeror was out of the office that day, or maybe even the following day? Perhaps the contracting parties could pre-empt problems such as these by agreeing, in advance and in writing, what type of communication is required, and when the acceptance will be valid.

Having said that the general law in contract states that acceptance is valid only when received by the offeror, an exception to this premise is now going to be examined. The above discussion concerned instances in which the medium of communication was, for all intents and purposes, instantaneous. What, however, is the law when communication is transmitted by post- where there is more delay between the sending of the acceptance, and the receiving of the acceptance by the other party? In the case of:

Adams v. *Lindsell* (1818)

The plaintiffs were woollen manufacturers based in Bromsgrove, Worcestershire. The defendants were wool dealers based in St Ives, Huntingdon. The defendants wrote to the plaintiffs on 2 September 1817, offering them a quantity of wool on specified terms. They required an answer 'in course of post'. This letter did not reach the plaintiffs until 5 September, because it had been misdirected by the defendants. The plaintiffs immediately sent back their acceptance to the offer which arrived at the defendants' premises on 9 September. The defendants were expecting a reply to their letter on 7 September, which would have been possible if they had correctly addressed and posted their original offer. As the defendants heard nothing on this date, the wool was sold to a third party on 8 September.

The court had to decide whether a contract for the sale of wool had been concluded between the plaintiffs and defendants before the wool had been sold to the third party. The court could have decided that the acceptance was valid,

1 when the letter of acceptance was in the hands of the offeror,
2 when the letter of acceptance was delivered to the defendant's premises,
3 when the letter of acceptance was placed into the post box.

The court determined that the last of these solutions should apply. At first, this may seem very strange, as a person may become a party to a contract without being aware of it! This could happen if a letter of acceptance was put into the post but never arrived at its destination, i.e. did not reach the offeror. The acceptance would be valid from the time it was placed into the post box, but the offeror, having never received the acceptance, would be unaware that the contract was, in fact, in existence.

The rule is not as harsh as it first seems. The 'postal rule' (as it has become known as) will only apply if two conditions are satisfied. The first of these conditions being that the letter of acceptance must be properly addressed, stamped and posted; and the second condition being that there was no method of communication stipulated by the offeror. Even though there may have been no method of communication stipulated by the offeror, the method of communication chosen by the offeree to inform the offeror of the offeree's acceptance must be a reasonable one in all the circumstances. This may mean that if an offer had been sent by telegram or fax, it could indicate that the offeror wished for a speedy reply. If, then, the offeree decided to send the acceptance by post, and the letter was lost and the offeror remained ignorant of the acceptance, what might the situation be if the offeror had sold the goods to a third party and the original offeree sued for breach of contract?

If the postal rule was applied, the acceptance would be valid the moment the letter was placed in the post box and a valid contract would have been formed. However, because the offeror may have been indicating that a speedy reply was required, the court may then decide that acceptance by post (a slower form of communication) is not a valid acceptance. There is authority to this effect in the case of *Quenerduaine* v. *Cole* (1883).

The offeror can be assured that a contract will not be formed, without having knowledge of the acceptance first, by stipulating that 'notice of an acceptance' *must* be received *in writing* in order to be valid. This was decided by the court in:

Holwell Securities Ltd v. *Hughes* (1974)

On 19 October 1971, the plaintiffs were granted an option, by the defendants, to purchase freehold property. The agreement included a clause which stated, 'the said option shall be exercisable by notice in writing to (the defendant) at any time within six months from the date hereof'. On 14 April 1972, the plaintiffs posted a letter to the defendants stating their intention to exercise the option. The letter, although correctly addressed and stamped, never reached the defendants. The plaintiffs then sought an order for specific performance from the court, or failing that, an award of damages. Specific performance is an order from the court which compels a party to carry out their contractual obligations. The plaintiffs could not succeed on either account unless there had been a breach of contract by the defendants. The plaintiffs claimed a contract had been concluded the moment their letter of acceptance of the option had been posted. This would have been an application of the postal rule. The court held, however, that there had been no acceptance, because the agreement had stated that, in order to exercise the option, the plaintiffs had to notify the defendants in writing, and this they had failed to do.

Waiver of communication

In situations where there is a unilateral contract, the need for the offeree to communicate acceptance is waived. The meeting of the conditions specified in the offer is all that is required to constitute an acceptance. An example of this can be found in the case of *Carlill* v. *Carbolic Smoke Ball Co.* where buying and using the smoke ball as directed constituted the acceptance.

A Short Summary of Offer and Acceptance

In order to establish whether a contract has come into existence between two or more parties, the courts will look at all the circumstances of the case, and apply the rules on offer and acceptance.

A checklist of important points:

1 During negotiations, distinguish between 'invitations to treat' and offers. Remember that invitations to treat have no legal consequences.
2 Offers must be firm and capable of unconditional acceptance. Enough information must be given to enable the offeree to agree to all the terms of the offer, otherwise the offeree may have to request further information. The offeror should respond to all such requests.
3 The offer can be withdrawn (revoked) any time before acceptance takes place.
4 Acceptance may be spoken, written, or inferred from the conduct of the parties concerned.
5 The acceptance must be unequivocal. If the offeree attempts to change any terms of the offer, this is not an acceptance, but a counter-offer, which must then be accepted or rejected by the original offeror.
6 The general rule is that the acceptance must be communicated to and received by the offeror before being valid.
7 An exception to this rule is the postal rule, which makes an acceptance by post valid as soon as it is posted. Remember that two conditions must exist before the postal rule will apply:
  (a) the letter must be properly addressed, stamped, and posted,
  (b) the offeror must not have stipulated any particular method of communication for the acceptance.

Remember also, that the postal rule will not apply if the offeror has stipulated that notice of the acceptance must be received in writing.

Consideration

Consideration is another principal ingredient the courts will look for in order to establish that there is a valid contract in existence. The courts look for an offer and acceptance in order to establish agreement between the parties. However, not all agreements are legally enforceable. If Mr Jones promised to mow the lawn, belonging to his next-door neighbour, Mr Smith, every week throughout summer, and then failed to carry out that promise, Mr Smith would not be able to sue Mr Jones for breach of contract. This is because:

1 There was no intention (of either party) to create legal relations (see next heading).
2 The agreement was merely a promise, and although there may be a moral obligation to carry out promises, there is no legal obligation to do so.

The presence of consideration distinguishes between 'promises', and 'legally enforceable agreements'. Consideration is often described as being the price of buying the promise from the other party to the contract. Take the following example:

Mr Jones promised to mow Mr Smith's lawn for him every week, and Mr Smith promised to pay Mr Jones for his services. There is no longer an empty promise to mow someone's lawn. The fact that the recipient is paying for the service, means that there is no longer just a moral obligation to mow the lawn, but there is a legal obligation as well.

This is now a normal commercial agreement (leaving aside the issue of intention to create legal relations) and if Mr Jones were to mow Mr Smith's lawn, and Mr Smith refused to pay, then Mr Jones would be able to sue Mr Smith for breach of contract.

In normal commercial contracts, the promise to supply goods and the promise to pay for those goods are all that are needed to demonstrate the presence of consideration, and render the contract legally enforceable.

There are two types of consideration:

- *Executory* consideration is the name given in the above example, where one promise is made in exchange for the other.
- *Executed* consideration arises where a promise is made in exchange for the performance of an act. An example of this situation is where a reward is offered for the return of missing goods. There is a promise of a reward,

once the missing goods have been returned. (This situation will be examined in more detail shortly, when offer and acceptance will be revisited.)

There can be a problem with consideration if the promises of the parties are not related, i.e. one is not made in exchange for the other. An example of this problem can be illustrated by examining the following case:

Re McArdle (1951)

A father, by his will, left the family house to his wife, during her lifetime, but after her death the house was to belong to his children. Whilst the wife was still living, one of her son's and his wife lived with her in the house. The wife made a number of improvements to the house. At a later stage the other children all signed a document which stated, 'in consideration of your carrying out certain alterations and improvements to the property, we hereby agree that the executors shall repay to you from the estate, when distributed, the sum of £488 in settlement of the amount spent on such improvements'.

When the time came, the money was not handed over to the wife and she sought to recover the sum through the courts. However, the court decided, that as the work had all been carried out prior to the promise of payment, this was not valid consideration. The work had not been undertaken in exchange for payment, but because the wife had wanted improvements made to the house. The work was therefore carried out independently to the later promise of payment. This was a case of *past consideration*, which is not a valid consideration, and so the wife's action failed.

Just to confuse issues further, there may be situations, in which the facts would appear to be very much like cases of past consideration, but in which the courts are prepared to find legally enforceable obligations The first case in which the courts took this view was that of:

Lampleigh v. Braithwait (1615)

Thomas Braithwait had killed a man and asked Lampleigh if he would seek a pardon from the King for him. Lampleigh endeavoured to procure the pardon from the King, which necessitated in him riding considerable distances at his own expense. He nevertheless obtained a pardon, whereupon Braithwait promised him the sum of £100 for his troubles. Braithwait failed to pay the money and Lampleigh sought to recover the £100 through the court.

The court decided that, although it appeared to be a case of past consideration, and therefore unenforceable, the later promise to pay was connected to the act, because the act had been instigated at the request of the defendant. Therefore, the promise and the act were to be considered as one transaction, the consideration was valid and the promise enforceable.

A similar approach was taken in the case of:

Re Casey's Patents, Stewart v. Casey (1892)

Stewart and Charlton were the joint owners of certain patents. They wrote to Casey, stating, 'In consideration of your services as the practical manager in working our patents, we hereby agree to give you one-third share of the patents.'

The case came before the court who had to decide if the letter was enforceable or not. In order for the court to find the agreement binding, Casey needed to have provided consideration for Stewart and Charlton's promise. Their representative argued that there was no valid consideration as Casey's actions had been carried out before the promise to award him a third-share of the patents. This meant that the consideration was past consideration and therefore invalid. The court, however, took a different view, by finding an implied understanding between the parties, that the services of Casey would be remunerated. This meant that the services Casey performed and the later promise of a third share in the patents, was considered to be linked as one transaction. This provided the necessary consideration for the promise to be binding.

The decisions in these cases have been followed more recently in the case of:

Pao On v. Lau Yiu Long (1980)

Lord Scarman explained the rule:

> An act done before the giving of a promise to make a payment or confer some other benefit can sometimes be consideration for the promise. The act must have been done at the promisor's request; the parties must have understood that the act was to be remunerated further, by payment or the conferment of some other benefit; and payment or the conferment of a benefit must have been legally enforceable had it been promised in advance.

Consideration must move from the promisee

This rule on consideration can be demonstrated by examining the case of:

Price v. *Easton* (1833)

Price owed the defendant the sum of £13. Price agreed to do some work for the defendant, who promised to pay the sum of money to the plaintiff. Price completed the work but the defendant refused to pay the money. The plaintiff sued in order to recover the money. The court held that the plaintiff was not entitled to sue the defendant for the money as consideration had not moved from the plaintiff to the defendant.

The value of consideration

The rule of law here is that consideration must be sufficient, though not necessarily adequate. The court is not concerned with whether a person has made a good bargain or not, or is getting value for money. This is for the parties to the contract to negotiate and decide for themselves. Therefore, it is up to the contracting parties to ensure the consideration is 'adequate'. The court is concerned with ensuring that the consideration provided is sufficient to support a legally binding agreement in the eyes of the law, i.e. that the consideration has *some* value.

This principle can be illustrated by examining the case of:

Midland Bank Trust Co Ltd v. *Green* (1981)

A father had granted his son an option to purchase his farm, but after a disagreement, sold the farm to his wife for £500. The farm had been valued at over £40,000. It was stated that the court would not enquire into the adequacy as long as the consideration was real. The House of Lords held that there was a valid contract between the husband and wife.

Existing contractual obligations and consideration

In the case of:

Stilk v. *Myrick* (1809)

A ship's crew had been employed to sail the ship from London to the Baltic and back. During the voyage some of the crew deserted the ship. The captain promised the remaining crew members that the deserters' wages would be split between them if they would work the ship home. On arrival at London, payment was refused. It was held that the crew could not enforce the captain's promise as they had provided no consideration for that promise. They were under an existing contractual obligation to work the ship home.

The promise would only have been enforceable if extra consideration had been provided. Extra consideration was found in the case of :

Hartley v. *Ponsonby* (1857)

The ship's captain promised the remaining crew extra money if they would work the ship home, after many of the crew had deserted. The sailors were refused the extra money once the ship was safely home. On this occasion, the sailors were awarded the extra money promised. The court found that:

(a) as so many of the crew had deserted, the task of working the ship home had become extremely dangerous, and

(b) the nature of their work was so different, from when the voyage had commenced, that by performing these extra, dangerous duties they had provided the extra consideration needed to form a legally enforceable agreement;

they were therefore entitled to the promised payment.

Performance of an existing contractual duty could not amount to good consideration. The performance had to go beyond the existing contractual duty in order to amount to good consideration. If the parties wished to vary the terms of a contract that was in force, then both parties to that contract would have to furnish consideration in order that the new contract terms were legally enforceable. Consider the following scenario. A and B have concluded a contract for the sale and supply of goods. A, now wishes to increase the price of those goods. If B agrees to pay a higher price than he is already obliged to pay, then A must furnish some extra consideration in order for B's promise to become legally binding.

It appears now that the courts may be willing to find consideration in some situations where extra consideration has not obviously been provided, in order to give legal effect to the parties' agreement. Consider the case of:

Williams v. *Roffey* (1990)

The defendants were the main contractors for the refurbishment of a block of flats. The plaintiffs were contracted as sub-contractors to undertake the carpentry work. They agreed a contract price of £20,000. The plaintiffs got into financial difficulties whilst carrying out their contractual obligations. This was partly due to the negotiated price being too low for the job. The defendants were worried that the plaintiffs would be unable to finish the carpentry work on the flats, or be extremely late in doing so. The reason for the defendants' concern was a penalty clause in the main contract which meant the defendants would be liable to pay a large sum of money in the event that the work was not completed on time. The defendants then promised the plaintiffs an extra £575 for each flat completed, which amounted to a total of £10,300. The plaintiffs then completed work on eight more flats. It then appeared to the plaintiffs that the defendants were not going to make the additional payments promised. The plaintiffs stopped their work and sued the defendants for the extra sums promised for work on the eight completed flats.

This would appear to be a case where there was no consideration to support the extra payment promised. The plaintiffs were already contractually bound to complete the work on the flats. However, the Court of Appeal decided otherwise and Glidewell LJ stated the following:

1 if A has entered into a contract with B to do work for, or to supply goods or services to, B in return for payment by B, and
2 at some stage before A has completely performed his obligations under the contract B has reason to doubt whether A will be able to, complete his side of the bargain, and
3 B thereupon promises A an additional payment in return for A's promise to perform his contractual obligations on time, and
4 as a result of giving his promise B obtains in practice a benefit, or obviates a disbenefit, and
5 B's promise is not given as a result of economic duress or fraud on the part of A, then
6 the benefit to B is capable of being consideration for B's promise, so that the promise will be legally binding.

The court found that the defendants obtained a benefit by not having to pay the penalty payment for late completion, and also by not having to find other contractors to complete the work.

Another area where consideration is deemed to be insufficient to support a legally binding agreement is that of part-payment of debts. A problem may arise where A owes money to B and promises to pay some of the debt if B agrees to forgo the sum remaining. If B accepts the sum proffered by A, can he then sue A for the balance remaining at a later date? In other words, is B's promise to accept a lesser sum than the sum due, binding on him?

The common law rule, is derived from Pinnel's case (1602), in which the following was stated:

> Payment of a lesser sum on the day in satisfaction of a greater cannot be any satisfaction for the whole, because it appears to the Judges that by no

possibility a lesser sum can be a satisfaction to the plaintiff for a greater sum. But the gift of a horse, hawk or robe, etc. in satisfaction is good. For it shall be intended that a horse, hawk or robe, etc. might be more beneficial to the plaintiff than the money in respect of some circumstance, or otherwise the plaintiff would not have accepted it in satisfaction . . . The payment and acceptance of parcel before the day in satisfaction of the whole would be a good satisfaction in regard to circumstance of time, for peradventure parcel of it before the day would be more beneficial to him than the whole at the day, and the value of satisfaction is not material.

This is not the case where part payment of a debt is made earlier than the date the debt is due. This is because there is some benefit to the creditor by receiving some payment at an earlier time. Also, if the creditor has agreed to accept goods, e.g. a car, in full settlement of the debt, instead of payment of the whole debt in money, he cannot go back at a later date and demand more money from the debtor, claiming that the car was not worth the amount of the debt. This is because the creditor must have derived some benefit from receiving a car or he would not have agreed to the transaction in the first place. The courts are not interested in the value of goods accepted in part-payment of a debt.

Therefore, there is insufficient consideration for part-payment of a debt unless the part-payment is made on an earlier date, at another place, or goods (of any value) are accepted in full settlement of the original debt.

In simple language, this means that if part payment of the debt is made on the day it is due, then the creditor can always demand payment for the remainder of the debt at a later date. The rule in Pinnel's case was followed in:

Foakes v. Beer (1884)

Mrs Foakes had obtained judgment against Dr Beer for the sum of £2,090. He asked Mrs Foakes for time to pay the debt. A written agreement was entered into, whereby payment of a lump sum of £500 was to be made immediately, followed by the remainder in instalments. It was also agreed that Mrs Foakes would not 'take any proceedings whatever on the judgment'. Neither party, at this time, had considered that interest is payable on judgment debts, calculated from the date of judgment. Dr Beer finally finished paying the judgment debt five years later. Mrs Foakes then claimed £360 interest on the debt. Dr Beer disputed her claim on the grounds of their former agreement. The court decided that the £360 must be paid, following the rule in Pinnel's case.

Another application of the rule can be seen in the more recent case of:

D&C Builders Ltd v. *Rees* (1966)

D&C Builders were owed £482 by the defendants for building work which had been completed. After several months, the plaintiffs were desperate for payment as they were suffering severe financial difficulties. The defendants were aware of the builders' financial problems, and Mrs Rees finally offered the sum of £300 in full satisfaction of the debt. She told them that if they did not accept this sum they would get nothing at all. The plaintiffs reluctantly accepted and banked the cheque for £300. They then sued for the balance and were successful.

Considering the decisions in the cases of *Foakes* v. *Beer* and *D&C Builders Ltd* v. *Rees*, it can be argued that the decision in the latter case was the correct decision to be arrived at, whereas the decision in the former case would appear to be very harsh. Indeed, the decision in *Foakes* v. *Beer* has often been criticized. The decision is obviously correct on the strict application of the common law in Pinnel's case, but whether the decision was fair, in all the circumstances, is debateable. The harshness of this decision has now partly been mitigated by the doctrine of promissory estoppel.

Promissory estoppel

Problems result from instances where part-payment of a debt is accepted by the creditor who is seeking to minimize his loss by accepting some payment towards the debt, albeit reluctantly, rather than receiving nothing at all. At a later date the creditor attempts to enforce the strict legal rights, i.e. the right to receive the remaining monies owed. When this happens, the debtor claims that the creditor agreed to accept the part-payment of the debt in full settlement and that because this was agreed upon, the debtor can no longer pursue the outstanding amount. As explained above, there is insufficient consideration for the agreement to accept part-payment to be legally enforceable, so the creditor's response would be that as the promise is not legally binding, the remainder of the debt must be paid.

In Chapter 1 it was explained how equity can step in to provide a fairer solution where the strict application of the common law fails to achieve a just result. Promissory estoppel is an example of this. Promissory estoppel is based on the doctrine of equitable estoppel, the principle of which Lord Cairns explained in the case of:

Hughes v. *Metropolitan Railway Co Ltd* (1877)

The appellant landlord served on the respondent tenants a six months' notice to repair the premises let. The respondents' failure to comply with the notice within the notice period, gave the appellants the right to terminate the lease. After the notice had been served, the parties commenced negotiations for the sale of the lease. The appellants then terminated the negotiations, which had been taking place for almost six months. The appellants then sought to forfeit the lease on the grounds that the respondents had failed to repair the premises. The House of Lords held that by entering into negotiations, the appellant had impliedly promised that no proceedings would be brought against the respondents for their failure to repair the premises within six months. The respondents had done nothing to the premises on reliance of this, and the six months' notice ran from the time when the negotiations broke down. The tenant was entitled to relief from the action to forfeit the lease.

Lord Cairns stated the following:

> It is the first principle upon which all Courts of Equity proceed, that if the parties who have entered into definite and distinct terms involving certain legal results – certain penalties or legal forfeiture – afterwards by their own act or with their own consent enter upon a course of negotiations which has the effect of leading one of the parties to suppose that the strict legal rights arising under the contract will not be enforced or will be kept in suspense, or held in abeyance, the person who otherwise might have enforced those rights will not be allowed to enforce them where it would be inequitable having regard to the dealings which have thus taken place between the parties.

The modern doctrine of promissory estoppel was founded in the case of:

Central London Property Trust Ltd v. *High Trees House Ltd* (1947)

The plaintiffs were owners of a block of flats in London which they let to the defendant tenants for a rent of £2500 per annum. After the outbreak of war in 1939 the defendants found it impossible to let all the flats because of the amount of people leaving London. The plaintiffs agreed to reduce the rent to £1250 per annum, but no time limit was fixed for this reduction. The defendants paid this reduced rent until after the war ended in 1945, by which time all the flats were once again full. The plaintiffs then sought to revert to the terms of the original agreement and restore the rent to £2500. They also wished to discover whether they could claim for the amount of the reduced rent for the duration of the war. Their submission being that they were entitled to do this as the agreement to reduce the rent was unsupported by consideration.

The court decided that the reduced rent was only intended to remain in force for the duration of the war and so the higher rent became payable from the end of the war. However, it was considered that any claim for the reduced rent during the war years would be unsuccessful. Denning J based his decision on the doctrine of equitable estoppel as stated in the case of *Hughes* v. *Metropolitan Railway Co Ltd* and stated, 'I prefer to apply the principle that a promise intended to be binding, intended to be acted on and in fact acted on, is binding so far as its terms properly apply'.

In the later case of *Combe* v. *Combe* (1951) Lord Denning explained the doctrine of promissory estoppel in the following terms:

> where one party has, by his words or conduct, made to the other a promise or assurance which was intended to affect the legal relations between them and to be acted on accordingly, then, once the other party has taken him at his word and acted on it, the one who gave the promise or assurance cannot afterwards be allowed to revert to the previous legal relations as if no such promise or assurance had been made by him, but he must accept their legal relations as if no such promise or assurance had been made by him, but he must accept their legal relations subject to the qualification which he himself has so introduced, even though it is not supported in point of law by any consideration, but only by his word.

It can be seen from the above that where a creditor has given a promise or assurance that part-payment will be accepted in full satisfaction of a debt, unsupported by consideration for that promise or assurance, equity may prevent the creditor from going back on the promise or assurance made to enforce the strict legal rights to recover the remainder of the debt. Remember, though, that equitable remedies are discretionary remedies and cannot be demanded as of right. Also equitable maxims apply, e.g. 'He who comes to equity must come with clean hands'. If then, there was any question of the debtor attempting to exert pressure on the creditor to accept less money by economic duress, promissory estoppel would not be allowed as a defence to the creditor's action in pursuing the remainder of the debt.

Strict principles must be followed for the doctrine of promissory estoppel to operate.

1 The promise must be clear, unequivocal and must be intended to affect legal relations and not merely be a gratuitous promise.
2 The promisee must have acted on the promise, or the position of the promisee must have changed in reliance of the promise
3 Promissory estoppel can only operate as a defence to an action by a party seeking to enforce strict legal rights.

From the above discussion on part-payment of debts, it can be seen that great care must be taken during negotiations between creditors and debtors. Creditors, when agreeing to accept a lesser sum in full satisfaction of a debt, may find themselves unsuccessful in obtaining the remainder of the debt at a later date. Debtors who seek to pressurize the creditor into accepting less than is due, may find their agreement voidable for economic duress.

Could the debtor argue that receiving some payment is of benefit to the creditor as the creditor will not then have to pursue a court action in order to obtain some of the money owed? This argument could be compared to that in the case of *Williams* v. *Roffey Bros.* which was a contract for goods and services In the recent case of Re Selectmove (1995) an attempt was made to apply this reasoning to a debt situation.

Re Selectmove

This case concerned taxes which were owed to the Inland Revenue by a company. The parties entered into negotiations about the payment of the taxes. The company offered to pay the taxes in instalments and the Collector of Taxes stated that the company would be contacted if this arrangement was unsatisfactory. The company commenced payment of the tax arrears in instalments. The company then heard that winding-up proceedings against the company would be commenced unless all arrears were paid immediately. The company attempted to avoid the winding-up petition by claiming that the Inland Revenue, by receiving the arrears in instalments, was receiving a benefit similar to that in *Williams* v. *Roffey Bros.* The benefit was in receiving the arrears without the cost of court proceedings to obtain the monies due.

The Court of Appeal distinguished the case of *Williams* v. *Roffey Bros.* on the grounds that *Williams* v. *Roffey* was a contract for goods and services, whilst the present case was one which concerned an existing obligation to pay a debt. This enabled the court to avoid following the precedent laid down in *Williams* v. *Roffey* and instead apply the rule in Pinnel's case as followed in *Foakes* v. *Beer*. This has had the effect of confining the precedent set in *Williams* v. *Roffey* to contracts for goods and services.

Another essential element in contract formation is:

Intention to create legal relations

The parties must intend their contract to be legally enforceable in order for the contract between them to be valid. Generally, the courts will look to see if the agreement entered into is a commercial agreement, or a domestic agreement. If the contract is concerned with a business deal, it will be regarded as a commercial agreement. There is a presumption, with commercial agreements, that the agreement is intended to be legally binding. There is no such presumption where the agreement is a social agreement between family or friends.

It is possible for the presumption to be rebutted (or disproved) but there must be very strong evidence for this to be accepted by the court. The two cases below give an example of this:

Jones v. *Vernon's Pools Ltd* (1938), *Appleson* v. *H. Littlewood Ltd* (1939)

In both cases, the plaintiffs tried to obtain monies which they claimed were winnings from a football pool. The defendants, in both cases, argued that as the football coupons contained the words, 'binding in honour only', they were not obliged to make the payments. The court agreed that the words were sufficient to rebut the presumption, and both plaintiffs failed in their action. The decisions in these cases may have been influenced by the types of contract that had been entered into, as these were unlike traditional commercial contracts.

Before the courts are willing to rebut the presumption in commercial contracts, that the contract is legally binding, the words used in the contract must be very clear, in order to avoid any ambiguity. The following case shows this clearly.

Edwards v. *Skyways Ltd* (1964)

The plaintiff, an airline pilot, was given three months notice of his redundancy by the defendants. As a member of the plaintiff's contributory pension scheme, the plaintiff was entitled to take his contributions out of the scheme, or to receive a pension, which would commence when he reached the age of fifty. Negotiations between the plaintiff's professional association and the defendants, resulted in an agreement being reached in which the plaintiff was to receive his contributions to date, plus an ex gratia payment equal to the contributions, if he chose this option. The plaintiff chose to withdraw his contributions and received the sum he had paid in. The defendants refused to pay the promised ex gratia payment.

The plaintiff sued the defendants for breach of contract. The defendants claimed that the agreement was not intended to create legal relations. The court found for the plaintiff. The court decided that as a business agreement had been made by the parties, there was a presumption that legal relations were intended. The onus was on the defendants to rebut the presumption. Because the presumption of intending to create legal relations is so strong, the defendants must take strong measures to rebut the presumption. This the defendants had failed to do. The court held that the words 'ex gratia', were insufficient to show lack of intention on the defendants' part. The defendants, therefore, were liable to make the payment to the plaintiff.

Capacity

Adults who are of sound mind have full capacity to contract. The law seeks to protect those who are incapacitated by placing restrictions on their ability to enter into contracts. Those who are incapacitated are minors, mentally disordered persons and drunks.

Minors

A minor is classified as an individual under the age of eighteen years by s1 Family Law Reform Act 1969. The law regarding contracts with minors seeks to protect the minor from unfair contracts, and the other contracting party in his dealings with minors. The general rule concerning these contracts, is that the minor will not be bound by a contract he has entered into, whereas the other party to the contract will be bound. As soon as the minor reaches the age of majority he will be able to ratify the contract if he wishes. The exceptions to this general rule are contracts for necessities, and employment contracts.

Employment contracts

Usually, it is in the interests of minors that employment contracts are binding, and so, minors will be bound by these contracts as long as they are of benefit to them.

Necessities

Where there is a contract for the sale of goods, s3(3) Sale of Goods Act 1979 states that, 'necessaries means goods suitable to the condition in life of the minor ... and to his actual requirements at the time of sale and delivery.' From this definition it can be seen that the circumstances of each individual will be taken into account in order to determine whether the contract is one of necessity or not. Contracts which provide the minor with food, drink, certain clothing and education will be classed as contracts for necessities.

Intoxicated Persons

If drunkenness prevents a person from understanding the terms on which a contract has been concluded, and the other contracting party is aware that the contract has been concluded with a person who is intoxicated, then the contract will be voidable. However, certain contracts may be binding. The Sale of Goods Act 1979, s3(2) provides that: 'Where necessaries are sold and delivered to a person who by reason of mental incapacity or drunkenness is incompetent to contract, he must pay a reasonable price for them.'

Mentally Incapacitated Persons

There are two categories of mentally disordered persons:

1 Under the Mental Health Act 1983, if a person is certified as being incapable of dealing with his property, by two medical practitioners, then the court will have control over the property. All contracts, except contracts for necessities, will be unenforceable.
2 Where there has been no certification, the contract will be voidable if the mentally disordered person did not understand the nature of the transaction and the other contracting party is aware of the person's disorder.

The Sale Of Goods Act 1979 s3(2), as stated above, applies, except where necessaries are purchased by the disordered person and the seller is unaware of his disorder. In this situation the seller may recover the full price for the goods.

Contract terms

Having considered the ingredients required in order for a valid contract to come into existence, the terms contained in a contract must next be examined. The two types of terms that will be discussed are express terms and implied terms.

Express Terms

As already discussed, contracts may be verbal contracts or written contracts.

In verbal contracts, the express terms are going to be those verbally agreed to by the parties. In the event of a dispute arising, the role of the judge is to determine what exactly those terms were, on the evidence before him. It can be seen from this that a written contract is always advisable.

Where all of the terms of a contract have been reduced to writing, these are express terms of the contract. In the event of a dispute between the parties, it is not for the judge to determine what those terms are, but to interpret the meaning of those terms within the contractual document.

Sometimes, during the negotiations to a contract, statements may be made which are not included in the written document. In this situation, in the event of a dispute arising, the court has to decide whether the statement made was a term of the contract or a mere representation. Where it is decided that the statement is a term of the contract, and that term has been breached, the innocent party may sue the other contracting party for breach of contract. Where it is decided that the statement is a mere representation, then, unless it can be shown that the statement was an actionable misrepresentation, the innocent party will have no cause of action.

It is vital then, to determine whether the statement was a term of the contract, or not. The courts attempt to discover the presumed intention of the parties by examining all the evidence available. As an aid to the courts' decision-making, the following guidelines are considered:

The importance of the statement

Bannerman v. *White* (1861)

During negotiations for the sale and purchase of hops, the buyer asked the seller if sulphur had been used in the treatment of the hops, saying that he would not bother asking the price of the hops if sulphur had been used. The seller assured the buyer that sulphur had not been used, and the sale went ahead. Later, it was discovered that sulphur had been used in the cultivation of a small percentage of the hops. The seller sued the buyer for the price of the hops and the buyer claimed he was justified in not complying with the obligations under the contract.

The only way in which the buyer would not be obliged to pay for the hops would be if the seller was in breach of contract. The seller would only be in breach of contract if the statement was in fact part of the contract, as claimed by the buyer. The buyer argued that the negotiations were part of one contractual transaction, and that the seller knew of the importance which the buyer attached to his enquiry.

The court found that the statement made by the seller was intended to be part of the contract and understood by both parties also to be part of the contract, as it was in response to such an important enquiry.

When was the statement made?

The statement must have been intended to be part of the contract and not merely part of the pre-contractual negotiations. In the case of:

Routledge v. *McKay* (1954)

The buyer and seller were negotiating the purchase and sale of a motorbike. On 23 October the seller stated that the motorbike was a 1942 model, after taking the information from the registration book. A written contract was drawn up on 30 October which made no reference to the year of registration. The buyer claimed for damages after the sale when it was discovered that the year of registration was actually 1930. The court held that the time lapse between the making of the statement and the written contract was too marked to be considered as one transaction.

Were the terms reduced to writing after the statement had been made?

If a written document follows statements which have been made, as in the case of *Routledge* v. *McKay*, the court must then decide whether the written document was intended by the parties to contain all the terms of the contract, or whether the contract was to be comprised of the written terms, and those terms which had been spoken as well. The court may well decide that only the terms which had been reduced to writing were intended by the parties to form the terms of the contract, as in *Routledge* v. *McKay*. However, in the case of:

Birch v. *Paramount Estates Ltd.* (1956)

The developers of an estate offered one of the new homes being built, to the buyers, promising that the property, 'would be as good as the show house'. After the plaintiff had agreed to buy the house, a written contract was drawn up which made no mention of the statement. The plaintiffs claimed damages when the house they had contracted to purchase, was not as good as the show house. The court awarded damages to be paid to the plaintiffs after finding that the statement was part of the contractual terms.

When negotiating the terms of a contract with another party, ensure that pre-contractual statements do not become part of the contract. Ensure that staff with the authority to negotiate contracts on behalf of the business, are aware that it is possible for statements to become contractually binding if care is not taken. When drafting the contract, ensure that all terms of the contract previously discussed, are covered. Any issues discussed during negotiations, but not intended as part of the contractual obligations, should be specifically excluded. It is preferable to have to conduct further negotiations before the terms are finally agreed to by both parties, than to enter into a legally binding agreement on unfavourable terms.

Had the maker of the statement special skill or knowledge?

Where the maker of the statement has special skill, or knowledge, that the representee does not have, the courts are more willing to find that the statement was a term of the contract and not a mere representation. This is demonstrated by the case of:

Schawel v. *Read* (1913)

The defendant was selling a stallion which the plaintiff was interested in buying. The plaintiff wanted the stallion for stud purposes and examined the horse to determine whether it was suitable or not. Before the plaintiff had finished the examination, the defendant interrupted by saying, 'You need not look for anything: the horse is perfectly sound'; whereupon the plaintiff ceased the examination. A price was agreed upon a few days later, and three weeks after that, the sale was concluded. The stallion proved to be unfit for stud purposes and the plaintiff sued the defendant for breach of contract.

The plaintiff claimed that the statement was a term of the contract and the defendant argued that the statement was a mere representation. The court held that the statement was a term of the contract. It was decided that the representation took place at the same time that the sale was concluded (it was one continuous transaction), and also that the defendant had special knowledge and skill regarding the condition of the horse, as he was the owner.

Particular care must be taken during the negotiation stage of contract formation, in order to reduce the risk of the other contracting party claiming that statements and claims were part of the contract and therefore legally binding.

The following two cases illustrate how the courts view other situations where one party has special knowledge, or, should have special knowledge of the subject matter of the contract. Firstly, in the case of:

Oscar Chess Ltd v. *Williams* (1957)

The plaintiffs were car dealers, who were selling a car to the defendant and taking the defendant's old car in part exchange. The part exchange value of the car depended on the age of the car. The defendant told the plaintiffs that the year of registration was 1948 (the year stated in the registration book from where the information had been obtained). The sale and part exchange was completed. Some months later the plaintiffs discovered the actual registration date was 1939. The plaintiffs sued the defendant for the difference in value between a 1948 model and a 1939 model.

The registration book had been altered before the defendant had bought the car and he had no knowledge of the alteration. The plaintiffs were car dealers and in a much better position to determine the truth as they had special knowledge and skill which the defendant did not possess. The court, surprisingly in this case, held that the statement was not a term of the contract.

Contrast this with the case of:

Dick Bentley Productions Ltd v. *Harold Smith (Motors) Ltd* (1965)

The plaintiffs informed the defendants that they wanted a 'well-vetted' car. The defendants informed the plaintiffs that they had such a car. Mr Bentley took the car for a test drive and was told that the car had only done 20,000 miles since being fitted with a replacement engine and gearbox. After Mr Bentley had bought the car, he experienced problems and also discovered the statement as to the mileage was incorrect. He sued the defendants for damages and succeeded. The court found that the statement was a term of the contract.

There is some academic disquiet as to how the decisions in these two cases can be reconciled. The Oscar Chess case had to be distinguished in order for the decision in the Dick Bentley case to be reached. Which decision will be followed in future cases will depend on the individual facts of the case; but, maybe these two decisions examined side by side, indicate the dangers that may be faced by persons in business making statements which are later found to be untrue.

So far, the discussion has focused on statements made by one contracting party, to the other contracting party, before the contract has been entered into, but which may become part of the contract.

These statements are express terms of the contract, as are terms which have been written down in a contract. However, not all terms of a contract are classified as express terms. There may also be implied terms of a contract. These implied terms are terms of the contract that have not necessarily been agreed to by the parties but are still part of the contract and are legally enforceable. The following discussion will look at different ways terms may be implied into a contract.

Terms implied by the courts

Generally speaking, the courts do not like implying terms into contracts. It has long been considered that persons who have reached the age of majority and are of sound mind can enter into a contract with another party on whatever terms those parties agree, without interference from third parties. It is in the interest of both parties to ensure that the terms of the contract are specified clearly and accurately reflect the intentions of the parties. The courts may intervene in certain situations where a term has not been expressly included into the contract, but one of the contracting parties claims that it is vital to the contract.

Terms implied by custom

The contracting parties may have assumed that their contract would be subject to commonly recognized rules of custom, associated with a particular trade, or locality. If there is sufficient evidence to support this assumption, the court will imply a term into the contract to give the rule of custom contractual effect. An example of this can be seen in the case of:

Hutton v. *Warren* (1836)

The plaintiff was the tenant, and the defendant the landlord of a farm. The landlord gave the tenant notice to quit. The landlord told the tenant that he was bound to continue to cultivate the farm, according to the custom of the country. On quitting his tenancy, in accordance with the notice to quit, the tenant claimed to be entitled to a fair allowance for seeds and his labour. The court found that it was, in fact, a well known custom for tenants to receive such an allowance. It was held that the custom had, by implication, been imported into the lease between the landlord and tenant, so the tenant was entitled to the allowance claimed.

Terms implied by the common law

Where a contract has been entered into, but does not reflect the obvious intentions of the parties, the courts may be prepared to imply a term to give effect to those intentions. The courts will only interfere if it is necessary to give business efficacy to the contract. The facts of the next case demonstrate this principle:

The Moorcock (1889)

The owner of the *Moorcock* had contracted with the owners of a wharf for the provision of mooring facilities. The ship was damaged, when moored, by rock at low tide. The owners of the *Moorcock* sued for compensation. The court implied a term into the contract between the parties that the wharf was suitable for mooring. The wharf owner was found liable for the damage because he had breached this implied term of the contract.

The court will only impose reasonable obligations on the parties to the contract. The term implied by the courts would be a term that both parties to the contract would have included had they given the matter any thought.

Terms implied by statute

Acts of Parliament may imply terms into contracts. One such example is the Sale of Goods Act 1979 (as amended by the Sale and Supply of Goods Act 1994) Several terms are implied by this Act. One example is given below. The other important terms are discussed in more detail in the next chapter.

Section 13 of the Act relates to the description of the goods, and states that: 'where there is a sale of goods by description, there is an implied condition that the goods correspond to that description.' A 'pure wool' jumper should be made from pure wool and nothing else, otherwise the seller is in breach of an implied condition of the contract, and the buyer will have a cause of action against the seller.

The different types of terms in a contract have now been explained, but the importance of these terms now needs to be considered. Some terms in a contract are deemed more important than others. A breach of an important term enables the innocent party to the contract to repudiate the contract, whereas a breach of a minor term in the contract merely allows the innocent party to the contract to claim damages. The important terms in contracts are known as 'conditions', whereas the minor terms are known as 'warranties'. The importance of the classification of conditions and warranties can be seen by considering the following two cases.

Poussard v. *Spiers & Pond* (1876)

An actress was contracted to play the lead role in an operetta from the beginning of its run. She failed to arrive for the play until a week after the commencement of the season, due to illness. The producers, meanwhile had had to engage another actress for the part, and refused her services. The court found that the term in the contract requiring her to be present for the start of the season was a condition of the contract, the breach of which entitled them to repudiate the contract.

Bettini v. *Gye* (1876)

A singer was engaged to sing at a series of concerts throughout the entire season. A term of the contract stated that he was required to attend for rehearsals six days before the first concert was to be performed. The singer arrived three days prior to the first engagement and the defendant sought to terminate the contract. The court decided that the requirement to attend for six days prior to the first concert was only a warranty, as the most important part of the contract was to perform at the concerts and this the singer was able to do. The defendant was unable to terminate the contract with the singer.

Conditions of a contract are said to be fundamental to the purpose of the contract, the breach of which strikes at the very root of the contract, destroying its purpose. There is one further category to be considered. This is known as an intermediate or innominate term.

Hong Kong Fir Shipping v. *Kawasaki Kisen Kaisha* (1962)

The defendants chartered a ship from the plaintiffs for a period of 24 months. A term of the charterparty was that the ship was to be 'in every way fitted for ordinary cargo service'. On the voyage to Osaka, the ship was delayed for five weeks due to problems with the engine. Fifteen weeks were then lost at Osaka because the engine room was undermanned and the engine staff were too incompetent to maintain the engine in good working order. The defendants repudiated the charter and the plaintiffs sued for breach of contract.

The court decided that the plaintiffs had breached the charterparty but that the breach did not entitle the defendants to treat the contract as terminated but were only entitled to claim damages. The court found that it was not always possible to classify terms of the contract as conditions or warranties from the outset of the contract. There were some breaches which would deprive the innocent party of substantially the whole benefit of the contract and some which would not. Some terms, therefore are neither conditions or warranties but are called innominate terms. If there was a breach of contract, the consequences of the breach would then be considered to discover whether the innocent party had been deprived of substantially the whole benefit which should have been received under the contract. If this was shown to be the case, then the innominate term would be treated in the same way as a condition and the innocent party would be at liberty to repudiate the contract. If it was found that the consequences of the breach were not sufficiently serious to deprive the other contracting party of the substantial benefit of the contract, then the term would be treated in the same way as a warranty and the innocent party would only have the right to claim damages for breach of contract. In the Hong Kong Fir case it was decided that the defendants had not been substantially deprived of the whole benefit of the contract as there was a considerable time left for the charter to run. Therefore they were not entitled to treat the contract as terminated. In the case of :

The Mihalis Angelos (1970)

A charterparty was entered into in May 1965. The ship was 'expected to be ready to load under this charter about 1 July 1965 at Haiphong'. The charter also gave the charterers the right to cancel the charter, 'should the ship not be ready to load on or before 20 July 1965'. The ship was not ready to load by 20 July and the charterers cancelled the charter. The owners of the vessel sued for breach of contract and claimed that the particular term of the contract giving the charterers the right to terminate the contract should be treated as an innominate term and the court should look to see the effect of the breach to discover whether the charterers had the right to terminate the contract. The court refused to take this approach, but instead took into account the wording of the charterparty, and the commercial necessity of knowing when a ship was 'expected to be ready to load'. It was decided that these clauses in charterparties were conditions of the contract, breach of which, would enable the charterer to terminate the charter. This decision was followed by the House of Lords in the case of *Bunge Corporation* v. *Tradax Export SA* (1981) which again concerned an 'expected readiness to load' clause.

The approach taken in the Hong Kong Fir case, of calling terms of the contract 'innominate terms', and then deciding when there is a breach, whether those terms are going to be treated as conditions or warranties, was discussed in later cases that came before the courts. One issue which was discussed, was the uncertainty faced by contracting parties of not knowing whether a breach of a particular term might result in an award of damages being made against the party in breach or whether, in fact, the innocent party could treat the contract as repudiated. Businesses need to be able to plan for as many contingencies as possible, which is not made easy if the result of a breach of contract cannot be predicted.

The courts will look to see whether the parties themselves have agreed whether contract terms are conditions or warranties. It may not be sufficient to state, 'it is a condition of the contract that . . . ' as 'condition' may not have the same meaning for both parties. If, however, this was followed by a statement of what the consequences would be in the event of a breach, e.g. 'breach of which will give the innocent party the right to terminate this contract', then the court can be certain what the parties intended. This does not mean, however, that the parties' own labelling of clauses will always be followed by the court. In,

Schuler AG v. *Wickman Tools Sales Ltd* (1973)

The defendants were given the sole selling rights of the plaintiff's panel presses. The contract stated that it was a condition of the contract that the defendants made visits to six named firms each week over a four and a half year period. Not all terms of the contract were labelled as either being conditions or warranties and Schuler's contention was that the contract could be terminated because the defendants failed to adhere to this condition. The House of Lords said that although the use of the word 'condition' would tend to mean that the contract could be terminated for breach, that on the facts of this case they could not agree with that particular interpretation. On the facts, 1400 visits were due to be made by the defendant, and a strict interpretation of that term as a condition would mean that there was a right to repudiate the contract if even one of these visits were missed. The House of Lords decided that there had been a material breach of contract which allowed Schuler to take action against the defendants under another clause in the contract.

In conclusion, if the term is reasonable and labelled by the contract then the courts are likely to give effect to the parties' intentions. When drafting contracts, thought should be given to the effect a breach may have on the business. If the other party to the contract is in breach, what effect may that have on the business of the innocent party?

Practical Considerations

Does the breach affect contractual obligations with any other party?

If a contract for the manufacture of certain machinery is breached by the manufacturer, the party who has ordered that machinery may be in the position where a contract to supply that machinery to a third party cannot be fulfilled. Damages may then become payable to the third party as compensation for that breach. A penalty clause in a contract with a third party may be invoked because of a late delivery. An order may be lost because a supplier has breached a contract to supply certain parts or goods under a contract. A breach of contract will often have a knock-on effect with other contracts and trading partners. Adverse effects on a business can be limited by preparing the terms on which the business trades as carefully as possible. A breach of contract can have devastating effects on a contracting party which could result in the collapse of the business. Liability for a breach of contract may be limited or excluded by the insertion of a suitable clause in the contract. The subject of exclusion or limitation clauses will be considered next.

Standard Form Contracts

Most businesses nowadays use standard form contracts. Standard form contracts are usually drawn up by lawyers for a company to use for every-day business. The standard form contract will normally contain a series of numbered clauses, drafted with contractual terms which are intended to favour the company. Problems have arisen in the past when both contracting parties have tried to contract on their standard form contracts which have contained different contractual terms. An example of this can be seen in the case of:

Butler Machine Tool Co. v. *Ex-Cell-O Corpn* (1979)

The sellers, in response to an enquiry from the buyers, sent a quotation for the sale of a machine tool at £75,535, delivery to be made in ten months time. The quotation included the sellers terms and conditions, which contained a price variation clause. There was also a statement that these terms and conditions would 'prevail over any terms and conditions in the buyer's order'. The buyer's reply was an order for the machine tool. The order was placed on the buyer's standard form contract which contained different terms and conditions, the most important difference being, that no price variation clause was included. At the bottom of the order was a tear-off slip acknowledging the receipt of the order, which stated, 'We accept your order on the terms and conditions stated thereon'. The sellers completed the acknowledgement slip and returned it, together with a letter stating that the buyer's order was entered into in accordance with the seller's quotation of 23 May. When the sellers delivered the machine tool to the buyers, the buyers were informed that the price had been increased by £2,892. The buyers refused to pay and the sellers brought an action for this increase, allowed for in the price variation clause.

The decision of this case will be given after considering the possible legal consequences of the parties' actions.

It was submitted at the start of this chapter that contracts are brought about by the 'agreement' of the parties. It could be argued here that:

1 There is in fact no agreement between the parties at all, and therefore no contract.

2 There could be a contract on terms taken from both parties standard form contracts.
3 There could be a contract on the seller's terms.
4 There could be a contract on the buyer's terms.

The problem with (1) is that the goods have been delivered and received. In cases such as this, the courts will often find that an acceptance to a party's offer had, in fact, been made. Acceptance may be found by the goods being delivered by the sellers. It could be said that the sellers have accepted the terms of the buyers contract, otherwise they would not have sent the goods to them. Alternatively, the buyers may be deemed to have accepted the terms of the sellers as they have accepted the goods. If the buyers had not wanted to contract on the seller's terms, they would have rejected the goods. A similar approach was taken by the courts in the case of:

Brogden v. *Metropolitan Railway* (1877)

This case was concerned with consignments of coal. The parties had shown themselves willing to be bound to a contract by delivering and accepting consignments of coal. In conclusion, it can be stated that where there is evidence that the parties intended to enter into a contractual relationship with each other and goods have exchanged hands, the courts will be willing to find a contract is in existence. The next problem for the court is to decide on whose terms the parties have contracted.

As regards the second possible solution, the courts are less willing to write the contract for the parties. The court will look at all the circumstances of the case, including what the contract states, the negotiations between the parties, and the parties conduct, in order to determine on whose terms the parties intended to contract.

Solutions (3) and (4) are both self explanatory.

The decision: The court analysed the communications between the parties in the following way.

The sellers, by their quotation, were making an 'offer' to sell the machine tool to the buyers for £75,535, subject to their price variation clause. The buyers, by their order, were making a 'counter-offer' to the sellers. The buyers' order was not an acceptance of the sellers' offer because all the terms of the offer were not unequivocally agreed to. A counter-offer has the effect of destroying the original offer and now becomes the offer which must be accepted or rejected. The sellers, by returning the buyers' acknowledgement slip, had accepted the buyers' counter-offer. The contract had therefore been concluded on the buyers' terms and conditions. The accompanying letter sent by the sellers was said to be merely confirmation of the description of the machine tool and the price.

The decision in this case has been criticized. It has been argued that the letter was not an agreement to all the terms of the counter-offer, as it was introducing new terms and that the sellers did not intend to contract on the buyers' terms and conditions.

This case illustrates the approach the courts may take when attempting to decide what contracting parties intended when a dispute arises and the contractual terms are uncertain. Care needs to be taken when entering into contractual relations with another party, that the terms are clear to both parties from the outset. Mistakes are costly, and a dispute may damage the commercial relationship irrevocably.

Standard form contracts have many advantages over tailor-made contracts. The contract will be drafted with the particular business of the organization in mind. The terms and conditions included will be suitable for all the day-to-day contractual transactions that are normally expected to take place.

Some advantages of standard form contracts are:

1 the organization saves money by not having to employ the services of a lawyer every time a contract is entered into;
2 the task of contracting with other parties can be delegated to junior members of staff, as there are no complex negotiations to take place – all the terms of the contract are in the standard form contract;
3 time is saved – as once the contract has been drawn up it can be used time and time again;
4 modern technology enables terms in a standard form contract to be amended or updated, quickly and cheaply;
5 senior members of the business are free to attend to other issues and bigger, more important contracts that need to be freely negotiated between the parties;
6 terms can be inserted into the contract which are more favourable to the business (subject to statutory restrictions – see 'excluding and limiting terms', later).

Some disadvantages of standard form contracts are:

1 problems may occur during the formation of a contract, if the person negotiating the contract for the business fails to deal with correspondence from the other contracting party correctly (as in *Butler Machine Tool Co.* v. *Ex-Cell-O Corpn*);
2 initial expenditure will be incurred in drafting the contract;

3 the other contracting party may also have a standard form contract on which terms they wish to deal; (putting in a clause stating, 'our terms and conditions will prevail', will not ensure the contract is actually entered into on those terms. If the other contracting party also includes the same clause, the most that can be achieved is a stalemate situation).

Most of the difficulties with standard form contracts can be overcome by careful drafting and a good quality assurance procedure being in place. If:

1 the initial drafting accurately reflects the requirements of the business, and
2 all staff are adequately trained, and
3 quality procedures are devised and followed,

then standard form contracts can prove an inexpensive and fast route to successfully contracting in business matters. Also, see Chapter 8 of this book, on quality management assurance procedures.

Excluding and Limiting Terms

These terms are often inserted into contracts for the purpose of:

1 excluding a party's liability for breach of contract or negligence, or
2 limiting liability to certain events in the contract, or
3 limiting the amount of compensation payable in the event of a breach of contract, or
4 an amalgam of any of the above.

Exclusion and limitation clauses are permitted to be included in contracts but Parliament has intervened to protect consumers from the effects of these clauses, by introducing the Unfair Contract Terms Act 1977. Before considering the effects of this legislation, the common law rules on incorporation and construction of exclusion clauses must be examined.

Incorporation of Exclusion Clauses by Signature

In order for an exclusion clause to be effective, the clause must form part of the contract. The party seeking to rely on the clause will have to show the court that the other party to the contract agreed to the clause before the contract was entered into. One way of showing this is where the contract contained an exclusion clause and was signed by the other party to the contract. An illustration of this can be seen in the following case.

L'Estrange v. *Graucob* (1934)

The plaintiff bought a vending machine from the defendant. The sales agreement included a clause which stated, 'any express or implied condition, statement or warranty, statutory or otherwise, is hereby excluded'. The plaintiff signed the agreement which she had not read. The vending machine often jammed and the plaintiff sued the defendant for breach of the implied terms that the goods were not fit for their purpose under the Sale of Goods Act 1893. The court held that in the absence of fraud or misrepresentation, a person who has signed a contract, whether they have read the contract or not, will be bound by the terms in that contract. The plaintiff therefore failed in her action. The first rule for all contracting parties at this stage is:

never sign a contractual document that has not been read carefully.

Incorporation of Exclusion Clauses by Giving Reasonable Notice – Exclusion Notices on Tickets

Exclusion clauses must be brought to the attention of the other contracting party before the contract is concluded. A clause can be brought to the other party's attention by giving that party reasonable notice of the clause. Whether the party has been given reasonable notice will depend on all the circumstances of the case.

Chapelton v. *Barry Urban District Council* (1940)

Mr Chapelton hired two deck chairs from the defendant council. He paid 2d per chair and received two tickets in return. He put the tickets into his pocket without having looked at them. One of the deck chairs collapsed causing Mr Chapelton injury. Mr Chapelton then sued the council for compensation for the injuries he received. The council claimed that their liability had been excluded by a clause on the reverse of the ticket which provided that, 'the council will not be liable for any accident or damage arising from hire of chairs'. The court held that the council could not rely on this exclusion clause. The ticket was not a contractual document, but a mere receipt for payment made. Reasonable notice of the exclusion clause had not been given.

Timing of notice

Reasonable notice of the exclusion clause must be given before the contract is concluded. This is shown by the case of:

Olley v. *Marlborough Court Ltd* (1949)

A husband and wife arrived at a hotel and paid at the reception desk. They went up to their room and on the back of the door was a notice which provided, 'The proprietors will not hold themselves responsible for articles lost or stolen unless handed to the manageress for safe custody'. Later the couple went out and took the key to the reception desk. The key was taken by a third party who stole a fur coat belonging to the plaintiff. The plaintiff sued for compensation and the defendant attempted to rely on the exclusion notice on the door. The court held that the contract was concluded at the reception desk and the exclusion notice was too late to be incorporated into the contract.

Incorporation of Exclusion Clauses by a Previous Course of Dealing

An exclusion clause may be incorporated by a previous course of dealing between the parties, as seen in the case of:

Spurling v. *Bradshaw* (1956)

Eight barrels of orange juice were delivered to the plaintiff by the defendant for storage. The plaintiff acknowledged receipt of the barrels. Included in the acknowledgement was a clause which excluded the plaintiff 'from any loss or damage occasioned by negligence, wrongful act or default'. When the defendant went to collect the barrels, he found that the barrels had been damaged and were empty. He refused to pay the plaintiff the storage charges that were due. The plaintiff sued for the payment and the defendant counter-claimed for negligence. The plaintiff claimed that the exclusion clause exempted him from liability. The defendant argued that the exclusion clause was too late to have been incorporated into the contract. It was admitted, however, that he had received such a document in previous dealings with the plaintiff although he had never actually read the document. The court held that the exclusion clause had become part of the contract through the previous course of dealings between the parties.

The courts are concerned that both parties to a contract are aware of the terms and conditions contained within that contract. The previous course of dealings between the parties must therefore be consistent. It will not be sufficient for a clause to be included sometimes but not others. The necessity for the course of dealings to be consistent is shown by the following case.

McCutcheon v. *David McBrayne Ltd* (1964)

The defendants operated a ferry service between the Scottish mainland and the islands. The plaintiff paid to have his car transported to the mainland. The ferry sank and the car had to be written off because of the damage suffered. The plaintiff sued and the defendants claimed their liability had been excluded by a clause contained in their printed terms and conditions. Reference to these terms and conditions was made both on the receipt and on a notice displayed in the office at the ferry terminal. A risk note which included the terms and conditions was not signed by the plaintiff. On some previous occasions a risk note had been signed, but not on every occasion. The court held that the defendants could not rely on the exclusion clause as there was no consistent course of dealing which would imply the exclusion clause into the contract between the parties.

Construction of Exclusion Clauses

Once it is has been established that the exclusion clause is part of the contract, it next needs to be established that the clause covers the breach complained of. The courts have long been hostile to exclusion clauses which will be construed against the party seeking to rely on the clause. The facts of the next case show how the courts will construe exclusion clauses.

Andrews Bros (Bournmouth) Ltd v. *Singer and Co.* (1934)

The contract between the parties stipulated that there was a contract for the purchase of 'new Singer cars'. An exclusion clause contained in the contract excluded liability for, 'all conditions, warranties and liabilities implied by statute, common law or otherwise'. The car dealers were sued for supplying a used car instead of a new car. The dealer sought to rely on the exclusion clause in the contract to avoid liability. The court, however, held that the exclusion clause did not cover the breach. The exclusion clause exempted liability for implied terms, whereas the term 'new Singer cars' was an express term of the contract. The dealer could not rely on the exclusion clause and was liable for the breach.

Incorporation of Unusual or Onerous Clauses

The more unusual or onerous a clause is, the greater the notice that must be given of the clause in order to incorporate the clause into the contract. This can be seen by the decision of the Court of Appeal in the case of:

Interfoto Picture Library v. *Stiletto Visual Programmes* (1988)

A contract was entered into for a number of photographic transparencies required for a presentation. A delivery note accompanied the forty-seven transparencies. Contained within the delivery note was a penalty clause which would take effect if the transparencies were returned after the specified date. The penalty amounted to £5 per transparency, per day. The transparencies were forgotten about and not used. They were returned about a month later. Interfoto sought to invoke the penalty clause and claimed over £3700 from the defendants for the late return of the transparencies. The Court of Appeal held that the clause could not be enforced. When the court were considering whether the clause had been incorporated into the contract or not they had to decide whether reasonable notice had been given of the clause or not. If reasonable notice had not been given, then the clause would not be part of the contract and Interfoto could not rely on the clause. The court held that because the terms of the penalty clause were so onerous then more steps should have been taken to bring it to the attention of the other contracting party. The terms were found to be onerous by comparing the terms of penalty clauses in similar businesses. Average charges for late returns were £3.50 per transparency per week and not the £5 per transparency per day charged by Interfoto. Although this clause was a penalty clause and not an exclusion clause the court said that the same rules for incorporation applied and reasonably sufficient notice had not been given in order for the clause to have been incorporated.

Excluding Liability for a Fundamental Breach of Contract

If one party wishes to exclude or limit their liability for breaching a contract, then the exclusion or limitation clause must be very clearly written. The clause must also cover the breach which has occurred and it must have been the intention of the parties to limit or exclude the breach. The next case demonstrates how liability for a fundamental breach of contract was excluded by one of the contracting parties.

Photo Production Ltd v. *Securicor Transport Ltd* (1980)

The defendants were providing a security patrol service for the plaintiffs. One night, the guard on duty started a fire which blazed out of control and burnt down the entire factory, causing some £615,000 worth of damage. The plaintifffs sued for their loss and Securicor sought to rely on an exclusion clause in their contract which exempted them specifically from 'liability for loss caused by fire . . .'. The Houes of Lords said that whether it was possible for an exemption clause to cover a fundamental breach of contract depended on the construction of the particular clause and whether it in fact covered the breach which had occurred. In this case the clause had been clearly drafted and covered the breach in question.

As stated at the beginning of the chapter, it has long been the premise that persons who have reached the age of majority and who are of sound mind can enter into contracts on whatever terms they choose. Individuals who enter into contracts with businesses are often contracting on the businesses' standard form contracts. There is an inequality of bargaining power in situations such as this, as both parties to the contract do not have an equal opportunity to negotiate the terms of the contract. The Unfair Contract Terms Act 1977 was introduced in order to redress the balance. The main beneficiaries under the Act are consumers, but businesses also have certain protection under the Act by making some clauses subject to a test of reasonableness.

The Act did not apply to the case of Photo Production, as it was not in force at the time. The Act was in force by the time the case reached the House of Lords and Lord Wilberforce explained what he thought the policy of the common law should be subsequent to the passing of the Unfair Contract Terms Act 1977. He said:

> It is significant that Parliament refrained from legislating over the whole field of contract. After this Act, in commercial matters generally, when the parties are not of unequal bargaining power, and when risks are normally borne by insurance, not only is the case for judicial intervention undemonstrated, but there is everything to be said, and this seems to have been Parliament's intention, for leaving the parties free to apportion the risks as they see fit and for respecting their decisions.

When deciding on the case it was noted by the court that the services of Securicor were inexpensive and that Photo Production were in the best position to insure their factory, so the clause would appear to be a reasonable clause to insert. Although the question of reasonableness did not arise in this instance, it demonstrates the issues the courts will take into consideration when applying the reasonableness test under the Unfair Contract Terms Act 1977. This will be discussed further in the section on the Unfair Contract Terms Act 1977.

Excluding Liability for Negligence

The discussion so far has centered on the exclusion of liability for a breach of contract. However, a contract may give rise to obligations under the tort of negligence as well as contractual obligations. The types of contracts where this is usually the case are works and materials contracts. It has already been stated that persons wishing to be excluded from liability under a contract must ensure that there is no ambiguity in the exclusion clause, otherwise it will be construed against them and the words used must be sufficiently clear to cover the breach. There is more difficulty if the contract gives rise to both contractual and tortious obligations and laibility for both of these is to be excluded. An example can be found in the case of:

Hollier v. *Rambler Motors* (1972)

The plaintiff contracted with the defendant garage to have his car towed to the garage for repairs. The contract contained the following clause: 'the company is not responsible for damage caused by fire to customers' cars on the premises'.

The plaintiff's car was totally destroyed by a fire caused by the defendants' negligence. The plaintiff brought an action against the defendants and they sought to rely on the exclusion clause. The court decided that the clause did not cover the breach. It was said that customers would assume that the clause only covered fires which were not caused through negligence. The clause was therefore ambiguous and the garage could not rely on it. If the garage wished to exclude liability for negligence they would need to use very clear words and exclude liability for negligence specifically.

The Unfair Contract Terms Act 1977

Schedule 1 to this act provides that certain types of contract are excluded from the operation of subsections 2, 3, 4 and 7. These include the following:

■ contracts of insurance;
■ contracts relating to the creation or transfer of interests in land;
■ contracts relating to the creation or transfer of rights or interests in intellectual property (copyrights, patents and trademarks);
■ international contracts as defined by the Act.

Always consult schedule 1 to determine whether the contract in question is covered by the Unfair Contract Terms Act 1977. The Act is further limited in its application by section 1(3). This section provides that ss.2–7 only apply to:

> ... business liability, that is liability for breach of obligations or duties arising (a) from things done or to be done by a person in the course of

a business . . . or (b) from the occupation of premises used for business purposes of the occupier. . . .

Effect of the Act on contractual liability

As stated previously, the Unfair Contract Terms Act either renders a clause ineffective, or subjects the clause to a test of reasonableness, depending on the parties to the contract.

Section 6 deals with contracts of sale (and hire-purchase). There are implied terms under section 12 of the Sale of Goods Act 1979 (and the corresponding legislation for contracts of hire-purchase) that the seller has the right to sell the goods at the time of the sale. Section 6(1) of the Unfair Contract Terms Act 1977 prevents a seller from excluding or restricting these implied terms. This applies irrespective of whether the seller is a private seller or a seller acting in the course of a business.

Section 6(2) provides that where the seller is dealing with a consumer, the seller cannot, by reference to any contract term, restrict or exclude any of the implied terms regarding the goods being of satisfactory quality, or fit for a particular purpose. Section 6(3) provides that where the seller is dealing with someone other than a consumer the implied terms referred to in s.6(2) can be restricted or excluded, but only in so far as the term satisfies the requirement of reasonableness.

Dealing as a consumer

Section 12 of the Act provides that a party to a contract 'deals as a consumer' in relation to another party if:

1 he neither makes the contract in the course of a business nor holds himself out as doing so; and
2 the other party does make the contract in the course of a business.

The goods passing under these contracts must be those ordinarily supplied for private use or consumption.

This shows that greater protection is given to persons contracting as a consumer, although the Act does give those who are acting in the course of a business some protection. In these circumstances, protection is limited to subjecting the exclusion clause to a test of reasonableness.

The reasonableness test

The reasonableness test is covered by s.11 of the Act. Section 11(1) provides that 'in relation to a contract term, the requirement of reasonableness . . . is that the term shall have been a fair and reasonable one to be included having regard to the circumstances which were, or ought reasonably to have been in the contemplation of the parties when the contract was made'. Schedule 2 of the Act specifies such matters which will be taken into account when determining whether a clause satisfies the requirement of reasonableness.

Schedule 2 – Guidelines for Application of Reasonableness Test

1 The strength of the bargaining positions of the parties relative to each other, taking into account (among other things) alternative means by which the customer's requirements could have been met. The courts will consider the size of the businesses of both parties and will also consider whether the customer was able to obtain the goods, which were the subject of the contract, readily from another source or whether the customer had a restricted market for the goods and was dealing with the other contracting party out of necessity.
2 Whether the customer received an inducement to agree to the term, or in accepting it had an opportunity of entering into a similar contract with other persons, but without having to accept a similar term.
3 Whether the customer knew or ought reasonably to have known of the existence and extent of the term (having regard, among other things, to any custom of the trade and any previous course of dealing between the parties).
4 Where the term excludes or restricts any relevant liability if some condition is not complied with, whether it was reasonable at the time of the contract to expect that compliance with that condition would be practicable.
5 Whether the goods were manufactured, processed or adapted to the special order of the customer.

Any of the above that are relevant may be taken into account for the purposes of determining whether a clause is reasonable. It is difficult to predict whether a clause will satisfy the reasonableness test in advance as the result will depend on the strength of the bargaining power between the parties to start with. Therefore the same clause inserted into a standard form contract may be reasonable against some parties and not others. Businesses who wish to rely on clauses that limit or exclude their liability to the other contracting party should ensure that such clauses are clearly and precisely written so that the other contracting party knows exactly the terms on which they are contracting.

3

The Tort of Negligence

The word tort is derived from a French word meaning a 'wrong'. Within the English legal system there are a number of torts; including those of defamation, trespass and nuisance, which are probably the most widely known. Tort is comprised within the area of civil law. Tort is concerned with compensating the victim for harm suffered rather than punishing the person responsible for causing that harm. The law provides the person 'wronged', the plaintiff, with the opportunity to bring a legal action in order to obtain compensation for the wrong which has been done to him.

As explained in the chapter on contract law, although the law concerning contracts affords a degree of consumer protection, such protection is limited in that it only applies to the actual purchaser of the defective product or service. The tort of negligence however, is independent of any contract being in existence between the parties. As a result it can afford a level of protection or a means of seeking redress to persons who may be harmed by a defective product or service, even though they do not enjoy the privity of contract between themselves and the supplier.

Actions in negligence are not restricted to injuries arising from defective products or services. Situations where accidental injuries arise, such as road traffic accidents, also often result in claims for compensation on the grounds of negligence. Such circumstances are unlikely to be within the remit of the quality practitioner. However, it should be noted that claims of negligence are the most commonly arising of all the torts for it covers a host of illnesses which may arise through working conditions, accidental injury situations or harm which may arise as a result of medical treatment. All of these areas could well be encompassed within a company's quality or combined quality/safety procedures. In addition to which, negligence also affords a degree of consumer protection, which is an area with which all quality assurance management systems should be concerned. An additional point to be borne in mind is that, it is not only negligent *acts* which need to be avoided. People involved with giving professional advice, such as lawyers, accountants and surveyors, need also to be aware of the potential liability they face for any negligent *statements* that they may make, or advice they may give.

As with all civil law, the burden of proof is on the plaintiff. The plaintiff must prove his case on the balance of probabilities. In cases of negligence the plaintiff has to be able to prove the other party was negligent. This is known as fault liability. To be successful in a negligence action, the plaintiff must show the following.

1 that the defendant owed the plaintiff a duty of care;
2 there was a breach of that duty of care; and
3 loss or injury arose directly as a result of that breach of duty of care.

The concept of a duty of care, was derived from case law rather, than from an Act of Parliament. Similarly exactly what constitutes a breach of that duty and whether or not injury arose directly because of the breach is also governed by case law. As a result the following pages contain an explanation of these terms together with the relevant cases from which they were derived, refined or put to the test.

Duty of Care

Since the case of *Donoghue* v. *Stevenson* (1932) a manufacturer can be sued in the tort of negligence for harm caused by a defect in their product. This case was the first in which a person was able to successfully sue the manufacturer in negligence. The facts of the case are as follows:

Mrs Donoghue's friend bought her a bottle of ginger beer when they were in a cafe in Paisley. After drinking most of the ginger beer, Mrs Donoghue poured out the rest onto some ice-cream, and the remains of a decomposed snail fell from the bottle. As a result of drinking the ginger beer, Mrs Donoghue alleged that she became very ill with gastro-enteritis. Mrs Donoghue wanted compensation for her pain and suffering. There was no contract between Mrs Donoghue and the cafe owner, as she was not the purchaser of the ginger beer, so there was no possible cause of action for breach of contract. The only course of action available to her was to sue the manufacturer of the drink in negligence.

As already mentioned in order to be successful in a negligence action, three points need to be proved:

1 A duty of care is owed to the plaintiff by the defendant.
2 The duty of care has been breached.
3 The breach of duty has caused the harm complained of.

Each point will be examined in turn.

Whether a Duty of Care is Owed

The problem Mrs Donoghue faced arose from the fact that, at that time, a manufacturer did not owe a duty of care to the ultimate consumer of his products.

Initially Mrs Donoghue was unsuccessful in her claim and in the subsequent appeal. However, a further appeal to the House of Lords resulted in a ruling which was to be one of the foundations of modern day negligence cases. The principle of law decided in this case was that a manufacturer of dangerous products owes a 'duty of care' to the ultimate consumer of that product. As previously stated, without a duty of care being owed, there can be no action in negligence before the courts. Perhaps one of the reasons for the judges arriving at the decision in this case was to encourage manufacturers to be more careful during the manufacturing process by making them more accountable financially, through the tort system, for mistakes or carelessness on their behalf.

It is easy to see from the above that consumable items would fall within the classification of 'dangerous products'. Any future instances of defective food or drink causing harm to a consumer would be covered by this precedent, in so far that the manufacturer would owe the injured consumer a duty of care. Providing that the consumer could prove that there had been a breach of duty by the manufacturer and it was the breach of duty which had caused the harm complained of, the consumer would be entitled to receive compensation from the manufacturer.

What would be the effect though of a non consumable item which could not be classified as a 'dangerous product' but nevertheless still caused harm to a consumer? In the case of *Donoghue* v. *Stevenson*, one of the judges in the case stated that the test for determining whether or not a duty of care was owed should be the 'neighbour test', He stated,

> You must take reasonable care to avoid acts or omissions which you can reasonably foresee would be likely to injure your neighbour. Who then in law is my neighbour? The answer seems to be ... persons who are so closely and directly affected by my act that I ought reasonably to have them in contemplation as being so affected when I am directing my mind to the acts or omissions which are called in question.

Under the 'narrow' test established in the case of *Donoghue* v. *Stevenson*, it would only be the manufacturers of 'dangerous products' who would owe a duty of care to their consumers, so it would appear to depend very much on what type of product has caused the harm complained of. However, if the words of the 'neighbour test' (the wide test) are examined again, it can be seen that the test could be applied to all manner of goods, as indeed it was in the later case of:

Grant v. Australian Knitting Mills (1946)

This was an appeal originating from the High Court of Australia to the Privy Council in England. It concerned a Dr Grant of Adelaide, South Australia who had bought several pairs of underpants from a store where the folded underpants were stacked on shelves. The underpants had not been individually wrapped and sealed by the manufacturer. Unfortunately, after wearing one of the pairs of underpants, Dr Grant developed dermatitis. The rash started on his ankle (he had purchased long johns) and then spread to other parts of his body. The dermatitis became so severe that Dr Grant was hospitalized for several months and actually feared for his life. He sought compensation from the manufacturers (the Australian Knitting Mills) for his injuries. It was argued by the defence that as underpants were not 'dangerous products', the court was not bound to follow the precedent in *Donoghue* v. *Stevenson*, and therefore the defendant – the Australian Knitting Mills – did not owe a duty of care to Dr Grant. However, by examining the words of Lord Atkin's 'neighbour test' it can be seen that a consumer is 'closely and directly affected' by the actions of a manufacturer. Also, a manufacturer 'ought reasonably to have them [the consumers] in contemplation as being so affected' when directing their minds to the acts or omissions which are called into question. Thus the Australian Knitting Mills *did* owe Mr Grant a duty of care. They should have taken reasonable steps to avoid sulphur being left in their products during production as it was reasonably foreseeable that omitting to do so would be likely to injure the 'neighbours' of the manufacturers (the 'neighbours' being the ultimate consumers of their products). The plaintiff, Dr Grant, by showing that sulphur had not been washed out of the underpants during the manufacturing process, was able to prove that the duty of care had been breached. It was also established that this breach of duty had caused the dermatitis complained of. Dr Grant was, therefore, able to receive compensation from the Australian Knitting Mills. The defendants had also tried to argue that as the underpants were not individually wrapped and sealed by themselves therefore it was not possible to show that the defendants were to blame for the condition of the underpants. Whereas the bottle of ginger beer in *Donoghue* v. *Stevenson* had been sealed with a metal top by the manufacturer, and this, it was claimed, rendered the present case completely different. It is the duty of the court however, to decide whether or not a duty of care is owed to the plaintiff by the defendant. A judge in the Australian Knitting Mills case stated the following:

> In the daily contacts of social and business life, human beings are thrown into, or place themselves in, an infinite variety of relations with their fellows; and the law can refer only to the standards of the reasonable man in order to determine whether any particular relation gives rise to a duty to take care, as between those who stand in that relation to each other. The grounds of action may be as various and manifold as human errancy and the conception of legal responsibility may develop in adaptation to altering social conditions and standards. The criteria of judgment must adjust and adapt itself to the changing circumstances of life. The categories of negligence are never closed.

What is clear from this part of the judge's speech and what needs to be emphasized, is that the courts are not only concerned with previous cases and the precedents that have been set, but are also open to consider all manner of situations that may arise in the future, not merely cases concerning manufacturers and their consumers.

The area of negligence as a tort has expanded over the years and the following are examples of situations where the courts have been prepared to find a duty of care between the plaintiff and the defendant exists.

Home Office v. *Dorset Yacht Co Ltd* (1970)

A group of borstal trainees escaped whilst on an outing due to the negligence of their guards and caused damage to yachts at the Dorset Yacht Co. The Home Office were held to be vicariously liable for the damage. *Vicarious liability* means that an employer – in this case the Home Office – is responsible for the torts of their employees, the guards concerned. It was decided that the Home Office owed a duty of care to the Yacht Co as it was reasonably foreseeable that damage may ensue if the guards were negligent and the borstal boys escaped. (Look back to the words of the 'neighbour test' to see how it applies to this situation.)

Remember that it is the 'duty of care' that is currently under discussion. Whatever business activity a person is engaged in, and whatever that person's role or position is within their organization they will owe a duty of care to someone. This may be just a few people or may be thousands of people depending on the type of activity being undertaken. Even as an individual, a duty of care will be owed to someone. As soon as a motorist sets off on a journey, a duty of care will be owed to all other road users (drivers and pedestrians). Teachers owe a duty of care to their pupils; doctors owe a duty of care to their patients and parents owe a duty of care to their children.

As can be seen by the examples given, a duty of care can arise from being involved in some form of activity, e.g. manufacturing, driving, or, when some form of responsibility has been assumed, e.g. parent/child, teacher/pupil.

The above examples have all involved either physical injury of some sort, or damage to property, but there can be occasions when damage to a person's mind can lead to liability. The type of damage being discussed here is not the kind of brain damage a person may suffer as the result of a road accident but a condition sometimes referred to as 'nervous shock', which a person may suffer from even though they themselves have not directly been involved in an accident. It may sometimes be family members of an injured victim, who although were not witness to the accident resulting in the injury, were witness to the aftermath within a short space of time. Another type of person who may suffer nervous shock is the 'rescuer'. This may be a passer-by at the time of an accident who stops to aid an accident victim, or indeed a professional such as a fireman, policeman, or ambulance driver who is called to the scene. Nervous shock is not merely grief that a person suffers at the loss of a loved one, or the anguish suffered by a helper at a scene of devastation or carnage, but is a recognized psychiatric illness resulting from being involved in some way with such an event.

The reader, at this stage, may be wondering what all this has to do with quality management and the individual at a place of work. Consider, firstly, all the persons to whom a duty of care may be owed. Using Lord Atkin's 'neighbour test' in *Donoghue* v. *Stevenson*, 'neighbours' are persons who are so closely and directly affected by an act that they should reasonably be in the contemplation of a person as being so affected, when directing one's mind to the acts or omissions which are called into question. Consider how many persons there may be that are closely and directly affected by actions at work, which fall in the category of 'neighbours'. It follows, if there is a negligent act which causes injury to a 'neighbour', that the injured person may succeed in an action for negligence.

Secondly, bear in mind the judgment in the case of *Grant* v. *Australian Knitting Mills* where it was stated that 'the categories of negligence are never closed'. Considerable thought should be given when constructing and reviewing quality procedures, as to how such acts may come about and what steps may be taken, with regard to working practices, training and supervision, to prevent such actions being undertaken. Given that the amount of compensation which may be payable could easily put a company or firm out of business, it is well worth considering the potential effects of negligence when constructing or reviewing a company's approach to ensuring quality.

After considering all the persons to whom a duty of care may be owed, thought must now be given to the standard of behaviour which must be achieved to avoid breaching that duty of care which is the second requirement for an action in negligence to succeed.

Breach of Duty of Care

In establishing whether or not a breach of duty of care has been committed, the law resorts to what is generally referred to as the *reasonable man test*. The meaning of the term 'reasonable man', is explained by the case of *Blyth* v. *Birmingham Waterworks Co.* (1856). In this case negligence was described as 'the omission to do something which a reasonable man, guided upon those considerations which ordinarily regulate the conduct of human affairs, would do, or doing something which a prudent and reasonable man would not do'. This means that the standard of behaviour which must be reached is that of the 'reasonable man'. This is an objective standard of behaviour which does not take into account individual capabilities or peculiarities. An examination of the following case illustrates the application of the reasonable man test.

Nettleship v. *Weston* (1971)

Mr Nettleship, a friend of Mrs Weston's husband, agreed to give Mrs Weston driving lessons. During the third lesson that Mr Nettleship had given her, Mrs Weston failed to manoeuvre the car correctly which resulted in the car hitting a lamppost. Mr Nettleship suffered injuries to his knee in the accident. He wanted to claim compensation for his injuries but the only way he could achieve this was to sue Mrs Weston in a negligence action. The first hurdle in proving negligence, was to show that Mrs Weston owed him a duty of care. This was easily shown as she was a car driver and so owed all other road users a duty of care. The second stage was to show that Mrs Weston had breached the duty of care owed. Mrs Weston's defence argued that she had not been negligent as she had driven as carefully as possible, because, as she was only a learner driver she could not have driven any better. The court accepted that Mrs Weston had driven to the best of her ability, but nevertheless decided that she had been negligent in the eyes of the law. The 'reasonable man' test, was an 'objective test', which meant that everyone was required to reach the same standard of driving regardless of whether they were learners or not. The standard of driving to be achieved, was that of the reasonably competent driver. The fact that she could not have attained that standard after only two lessons was immaterial. When judges are applying the law to a certain set of facts in the cases before them, they may also take other factors into account as well. If Mrs Weston had not been found to be negligent then Mr Nettleship would not have been able to claim compensation for an injury caused by another person. Furthermore, because Mrs Weston was insured, she would not be paying the compensation out of her own money, so that she would not be penalized in any way. In fact one of the judges told her that she was not 'morally' to blame, but only 'legally' to blame.

At first glance this case would seem to be specific to road traffic accidents, however, consider the similarities with industrial or service applications. In essence it concerns training, and the assurance that the person being trained has reached a sufficient state of competence before being permitted to carry out the task for which they are being trained. It is the accepted system within the United Kingdom that drivers of motor vehicles undertake their training on the public highways with all the inherent risks which that involves. However, factories with moving machinery, kitchens, waiting staff, attendants at theme parks, indeed just about every manufacturing or service organization at some point, usually from necessity, allows partially trained or newly trained personnel into situations where their negligence could well be harmful to others. What needs to be borne in mind as a result of the *Nettleship* v. *Weston* case is that lack of expertise is not an acceptable defence. As a result, the quality of training procedures should be such that any operative is brought up to the necessary standard in the course of their training and that supervision, recording and monitoring of that training should be designed to ensure that certain standards have been met. There should also be procedures in place to ensure that training is updated or reviewed on a regular basis.

The standard of care will be judged by the knowledge of the defendant at the time the incident happened.

Skilled persons

If a person holds himself out as having a particular skill he will have to exercise the level of skill normally possessed by persons engaged in that type of work. Therefore, a consultant will have to reach the standard of the reasonably competent consultant, an electrician will have to reach the standard of the reasonably competent electrician and a pilot must reach the standard of the reasonably competent pilot.

The standard of behaviour for us all has therefore been set by the standard of 'the reasonable man', when considering whether a person has been negligent or not.

Breach of Duty

The next hurdle to be overcome is that the breach of duty must have caused the harm complained of. This is best explained by the case of :

Barnett v. *The Chelsea and Kensington Hospital Management Committee* (1968)

Mr Barnett was a night watchman who was on duty with two colleagues. They had all not long finished a tea break, when Mr Barnett complained of severe stomach pains. His colleagues took him to the Chelsea and Kensington Hospital. The registrar on duty failed to examine him, advising him instead to visit his own doctor the following morning. Unfortunately Mr Barnett died during the night. As a result Mr Barnett's wife brought an action against the Hospital Management Committee, claiming that the registrar had been negligent in failing to examine her husband whilst he was at the hospital.

To succeed in the action, Mrs Barnett had first to show that the registrar owed her husband a duty of care. This she was able to do, as there is a recognized duty of care between a doctor and his patient. The registrar owed a duty of care to all his patients, and Mr Barnett automatically became a patient by arriving at the hospital complaining of feeling unwell. Therefore, the registrar owed Mr Barnett a duty of care. The second stage in the negligence action, was to show that the registrar had breached his duty of care by failing to reach the standard of the reasonably competent registrar. In this instance, the court decided that the duty of care had been breached, as any reasonably competent registrar would have examined a patient arriving at the hospital in the same circumstances. The final stage then, was the need to prove to the court that it was this breach of duty of care which had caused Mr Barnett's death. Mrs Barnett was unable to show this, as a post mortem examination, revealed that Mr Barnett had been suffering the effects of arsenic poisoning. Even if the registrar had examined him and admitted him to the hospital, Mr Barnett would have died anyway.

The failure to examine was not the cause of death, therefore the registrar was not negligent and Mrs Barnett's claim was unsuccessful.

It can be seen from the above examples that many people working in an industrial, commercial, or the retail environment will owe many other people a duty of care. The breach of that duty may result in legal action being taken, where the negligent person may find themselves liable to pay compensation to the injured party. However, in instances where a danger of a particular kind could not reasonably have been foreseen by the defendant, he will not have acted negligently. A person need not take precautions against unforeseen events – the reasonable man will need only to guard against risks that were known at the time. This point is highlighted by the case of:

Roe v. *Minister of Health* (1954)

Mr Roe went into hospital to undergo an operation. He was injected with an anaesthetic which had been contaminated by disinfectant. The syringe had been stored in phenol which had seeped through invisible cracks in the syringe. As a result of this Mr Roe was left paralysed. He sued for compensation for the harm caused to him. The first hurdle was to show that he was owed a duty of care. This was possible as it has been established that doctors owe their patients a duty of care. The next hurdle was to establish that the duty of care has been breached. Mr Roe failed to establish this, as it was unknown at the time that syringes could be contaminated in this way. He therefore failed in his negligence action and was unable to claim compensation.

If a reasonable man foresees harm, then he should conduct himself in such a manner to avoid such harm. He will then be behaving as a reasonable man should, and will not be negligent. A point to be made here is that this case should not be taken as an indication that ignorance of the possibility of an event is a defence against an action brought in negligence. Indeed it is a maxim of English law that ignorance is no excuse. Hence the failure, for example, to conduct risk assessment should not be seen as an advantage it that no potential risks were identified. On the contrary not only would such a failure constitute a poor attitude to quality awareness it may well, dependent upon what subsequently transpired, result in actions for negligence arising, or even for events taking place which could result in prosecutions being brought under the Health and Safety at Work Act 1974. Hence it is the likelihood of an event occurring which will be one of the most important issues to be considered by the courts when determining whether a person has breached their duty of care. An illustration of this can be found in the case of:

Bolton v. *Stone* (1951)

Miss Stone, the plaintiff, sued the members and committee of a cricket club under the name of Bolton. The cricket club appealed against the decision of the Court of Appeal who found the cricket club liable. The cricket club are the appellants here and Miss Stone is the respondent. The facts are as follows; the plaintiff, who was walking along a road, was struck by a cricket ball, which had been hit by a batsman at the defendants' cricket ground. It was a remarkable shot by the batsman as it cleared a seventeen foot high fence around the cricket ground and it was only about the sixth time the ball had been hit out of the ground in thirty years. The risk of such an accident occurring was therefore foreseeable, but the chances of the accident materializing were very small. The House of Lords held that the defendants had not been negligent in ignoring the risk as it was reasonable in all the circumstances of the case to ignore such a small risk. So, the likelihood of an event happening must be weighed against the seriousness of the risk and the cost and practicability of avoiding that risk. However, in instances where the consequences of the risk of harm are serious, every possible step needs to taken to avoid an accident occurring. A principle which was illustrated by the case of:

Paris v. *Stepney Borough Council* (1951)

This was a case involving the liability of an employer to an employee. The employers, the defendants in question, were aware that their employee, the plaintiff Mr Paris, was blind in one eye. During the course of the plaintiff's employment, a splinter of metal entered his good eye, rendering him totally blind. Mr Paris alleged that his employers had been negligent in failing to provide him with safety goggles, even though no other employees, doing the same work, were provided with goggles. The House of Lords in their judgment agreed with Mr Paris and held that the employers were liable for the injuries he sustained. As the duty of care, which was owed by an employer to an employee, was to each individual employee and as they knew that Mr Paris was blind in one eye, they also should have been aware that any injury sustained by Mr Paris would have potentially more serious consequences than it would have had if it had been sustained by an employee with sight in both eyes. It was decided that the employer's duty in this case included taking extra precautions to avoid the possibility of this type of injury occurring.

From what has been explained so far it would seem that employers, firms and businesses must try to guard against eventualities which are within their control. However, there are some occasions where defendants may be justified in taking risks during a particular activity because of the social utility of the exercise, as shown by the case below:

Watt v. Hertfordshire County Council (1954)

Firemen were called out to an accident where a woman was trapped underneath a lorry. A heavy lifting jack was needed in order to free the woman, but a specially designed vehicle for carrying the jack was not available. The jack was carried on an ordinary lorry supported by three firemen. One of the firemen was injured when the jack slipped. He claimed his employers were liable for his injuries. The court however, held that the employers were not at fault by risking injury to a fireman in order to reach an emergency quickly. Lord Denning said:

> It is well settled that in measuring due care you must balance the risk against the measures necessary to eliminate the risk. To that proposition there ought to be added this: you must balance the risk against the end to be achieved . . . the saving of life or limb justifies taking considerable risk.

He further added that if the accident had taken place within a commercial activity and there had been no danger to life or limb then the plaintiff would have succeeded. So, the saving of life or limb justifies taking considerable risk, but that does not mean that the taking of any risk is justified. Members of rescue services who fail to observe a red light may well find themselves liable in a negligence action for causing harm to an individual whilst on the way to help another individual who is the victim of an accident.

Not all risks are avoidable and some risks can only be minimized or eliminated at an exorbitant cost. It may not be practicable to eliminate these risks as the high cost of eliminating or minimizing the risk may outweigh the slight chance of the risk occurring. The following is an example of this situation.

Latimer v. AEC Ltd (1953)

Mr Latimer was injured when he slipped on the factory floor. Water had found its way into the factory after some heavy rainfall and had mixed with oil, rendering the floor very slippery. The employers had spread sawdust over the floor, but there was a small area of the floor which remained untreated, as their supply of sawdust was exhausted. The only other steps available to the employers would have been to close down the factory, which would have resulted in a considerable loss to the business. The court held that the defendants had behaved as reasonable employers and had taken all reasonable steps to minimize the risk of accidents occurring, short of closing down the factory. If the risk to the employees had been very great, then it may have been necessary for the employers to have taken this course of action, but in all the circumstances of this particular case, the cost and practicability far outweighed the risk of an accident occurring and so the employers were justified in keeping the factory open.

It can be seen from the above cases that the courts will take into consideration all the circumstances of the case when deciding whether a person has behaved reasonably in a negligence action.

Who Does Negligence Concern?

In the workplace, a person may have to guard against injuring:

- fellow employees;
- consumers of the product being produced;
- customers being supplied with goods which have been assembled;
- persons involved with the transportation or storing of such goods;
- persons involved with the fitting or installation of such goods;
- persons involved with the maintenance or repair of such goods.

In instances where there is no product being produced, an employee has to take reasonable care to ensure that no injury results to persons in their care. These persons could include:

- patients in a hospital, hospice, doctor's surgery, dentist's surgery or any other place where a person is receiving treatment at the hands of another;
- pupils in a school; college, university, or any other teaching or training establishment;
- clients in a hairdressing or beauty salon;
- clients in health farms and sporting establishments;
- clients in hotels, restaurants and cafes . . .

The above list, which no doubt could have further additions, demonstrates that whatever type of business a person is engaged in, they will owe a duty of care to someone. Making people aware of the legal consequences of their actions may encourage them to behave more responsibly in the workplace, and to consider those persons working alongside them. Also, to consider other persons who may be affected by their actions. It may well be worthwhile for the reader to consider for a moment the training procedures in his or her own place of employment. In the vast majority of cases no doubt the technical aspects of the job are covered, as will be the interpersonal skills when dealing with customers and the immediate safety requirements. But how far does the latter extend? For example: those people who work with machinery, the danger to themselves and those in their immediate vicinity will no doubt be explained, but does the training consider the duty of care which needs to be extended to all those who may suffer harm as a result of that particular operation. Does, in fact, the term duty of care and what it means actually form part of the training programme? Given the number of people within the categories above, who could be injured as a result of the operation of any particular item of machinery it would seem apparent that the responsibilities with regard to negligence should form part of any training programme. This of course equally applies to the service industries.

General or Common Practice

If, within a particular field, there is a general or common practice procedure in place which other persons, in a similar situation, would follow, and this practice is adopted, it would be unlikely that a court would find the actions to have been negligent. Failure to follow a common practice, conversely, would not automatically indicate a finding of negligence. In the case of:

Ward v. *The Ritz Hotel (London) Ltd* (1992)

The hotel had failed to comply with the British Standards Institution's recommendation on the height of a balustrade on a balcony. The court found that this was strong evidence of negligence but was not conclusive proof.

Where the common practice is adopted, a finding of negligence would be likely if it was discovered that the system had not been reviewed or updated. There is an obligation to keep up to date with knowledge and skills. Practices may need to be altered or adapted in the light of new knowledge. Given that review procedures should be an integral part of a quality assurance management system this is one example of where the correct formulation and function of such a system can be beneficial in helping a company adhere to the requirements of the law.

The Burden of Proof – res ipsa loquitur

There may be occasions when an accident results from circumstances in which accidents would not normally happen unless someone was found to have been negligent. In situations such as this, it would be very onerous on the plaintiff to have to prove the defendant had been negligent, especially when there is no evidence to show how the accident had occurred. In instances such as this, it may be possible for res ipsa loquitur to be raised. *Res ipsa loquitur* means 'the thing speaks for itself'. The effect of this being invoked is that an inference is raised that there must have been an act of negligence in order for the accident to have occurred. The defendant now has to rebut this presumption by proving that he was not negligent. The evidential burden is reversed by *res ipsa loquitur*, so instead of the plaintiff having to prove that the defendant was negligent, the defendant must prove that he was not negligent, or raise another plausible explanation, not involving negligence, for the occurrence of the accident. This doctrine can only be invoked in certain situations where all three of the following conditions are present:

1 The accident happens under circumstances in which accidents would not normally occur without negligence. This principle is explained by looking at the case of:

Scott v. *London and St Katherine Docks Co.* (1865)

The plaintiff was working in the defendants' warehouse when several large bags of sugar fell onto him from a hoist. In the first hearing, the judge found for the defendants as there was no evidence of negligence against them. In the new trial ordered by the Court of Appeal, a judge said:

> There must be reasonable evidence of negligence. But where the thing is shown to be under the management of the defendant or his servants, and the accident is such as in the ordinary course of things does not happen if those who have the management use proper care, it affords reasonable evidence, in the absence of explanation by the defendants, that the accident arose from want of care.

2 The thing causing the damage must have been under the exclusive control of the defendant or those for whom he is responsible.

Easson v. *London & North Eastern Railway Co.* (1944)

A small boy fell through the door of a train during a journey from Edinburgh to London. The accident happened when the train was several miles from the last station stop. The court held that the doctrine of res ipsa loquitur did not apply as it could not be shown that the door of the train was under the sole control of the railway company or their servants. It was possible that the door had been interfered with by any of the passengers on board.

3 The cause of the accident must be unknown. If all of the facts of the accident are known, then res ipsa loquitur will not apply and the only question to be answered is whether, on the facts, negligence by the defendant can be inferred.

Barkway v. *South Wales Transport Co. Ltd* (1948)

The plaintiff was injured when a tyre burst on the bus he was travelling on, causing the bus to crash. It was discovered that the tyre wall was defective. Because these facts were known, res ipsa loquitur did not apply. The defendants were found to have been negligent for failing to instruct their drivers to report any heavy blows to tyres on their buses.

Defences, Exclusions and Limitations to Negligence Actions

A company's approach to quality should be synonymous with its approach to its legal obligations, and for its own protection both commercial as well as legal, adopting the quality ethos of 'right first time' would appear to provide the ideal solution. However, good quality procedures regularly reviewed may not prove to be an all-encompassing panacea with regard to preventing actions for negligence arising (bearing in mind that 'the categories of negligence are never closed' – Australian Knitting Mills case). A service that such procedures may serve, if not prevent, may well be to assist with the defence against a potential action. Remember what needs to be established by a plaintiff:

1 that the defendant owed a duty of care to the plaintiff;
2 that a breach of that duty of care occurred; and
3 the breach of duty resulted in the damage complained of.

The plaintiff will succeed in a claim for negligence when the above is established unless the defendant has a defence to the claim. So what are the possible defences?

Defences

Volenti non fit injuria

The first defence to be dealt with is that referred to as *volenti non fit injuria* (that to which a man consents cannot be considered an injury). If a person voluntarily waives a right then he cannot later attempt to enforce that right in a court of law; as a result, if a person expressly or impliedly agrees to an act, then that act is not actionable as a tort. In other words, the plaintiff is agreeing to grant the defendant exemption from liability for causing an unreasonable risk of harm, in circumstances where the plaintiff has knowledge of the type and extent of that risk.

There is a slight difference in the operation of the defence of *volenti non fit injuria*, depending on the circumstances in which the defence is raised. In circumstances concerning intentional torts, the defence serves to act as consent. An example of the operation of the defence can be seen in the case of two boxers fighting in the ring. They are punching each other intentionally, and if one of the boxers decided to sue the other, for battery, the defence of *volenti* could successfully be raised, as by participating in the fight, both boxers are consenting to being hit by the other party.

In the case of negligent acts, the defence is more difficult for the defendant to raise. The effect here is that the plaintiff consents to exempt the defendant from the duty of care that he would otherwise owe him.

The agreement of the plaintiff, to grant the defendant exemption from liability, must be given voluntarily. Therefore, the courts will look to determine whether the plaintiff exercised a choice when consenting to the act of the defendant, in cases where the defendant raises the defence of *volenti*.

In the case of *Bowater* v. *Rowley Regis Corporation* (1944), Mr Lord Justice Scott stated:

> A man cannot be said to be truly willing unless he is in a position to choose freely, and freedom of choice predicates not only full knowledge of the circumstances on which the exercise of choice is conditioned, so that he may be able to choose wisely, but the absence from his mind of any feeling of constraint so that nothing shall interfere with the freedom of his will.

If the plaintiff is coerced into exempting the defendant from liability then the defendant will be precluded from successfully raising the defence of volenti as the plaintiff will not have exercised a genuine freedom of choice.

Similarly in the case of *Nettleship* v. *Weston* (1971), where Mr Nettleship was injured whilst giving Mrs Weston driving lessons, it was held that Mrs

Weston's defence of *volenti* failed. Mrs Weston claimed that by agreeing to give her lessons, Mr Nettleship had agreed to the risk of an accident occurring. The court, however, decided that Mr Nettleship had not consented to the risk of injury, as he had checked the insurance on the car in order to determine whether, as a passenger, he was covered by the terms of the insurance policy.

The knowledge of the plaintiff extends not only to the existence of the risk, but also to the nature and extent of the risk. In this case, it is the plaintiff's subjective knowledge that is taken into account by the court when determining whether the defence of *volenti* will succeed. The effect of this subjectivity, is that if the plaintiff *should* be aware of all the facts relating to the nature and extent of the risk, but is not aware of such factors, then the defence of *volenti* will not succeed against him. Such a situation would arise, if for example, the plaintiff was drunk at the time of the incident taking place, and as a result could not be said to be unaware of the extent of the risks. In such a case the plaintiff could not be considered to be *volenti*.

In situations where the parties concerned have expressly reached agreement that the plaintiff will voluntarily assume the risk of injury, the agreement may fall subject to the provisions of the Unfair Contract Terms Act 1977 (UCTA 1977). This Act was considered in more detail in Chapter 2, but suffice it to state here that in some cases it is impossible within the terms of the Act to exclude liability for negligent acts leading to death or personal injuries. The defendant, in these circumstances, will not be able to commit a negligent act, causing physical harm to the plaintiff, and then raise the defence of volenti, as the Act forbids the operation of such a defence.

Volenti non fit injuria: special cases

Employment cases   In the area of employers' liability, it used to be the case that employees rarely succeeded in negligence actions against their employers, for industrial injuries received at work. Many of these claims were defeated by the defence of *volenti non fit injuria*. The courts, at that time, took the view that employees knew the risks associated with going to work and that if they were injured at work then they had agreed to the risk of injury by turning up for work. Thankfully, the courts adopt a less rigid approach these days and there is also a wealth of statutory protection in force which makes employers reasonably responsible for the safety of their employees. The case in which the court recognized that there was an economic need to attend work, and took a more modern view of the employer/employee relationship, was:

Smith v. Baker (1891)

The plaintiff, Mr Smith, was injured whilst working on the construction of a railway. A crane carrying rock constantly swung its load across the site where the plaintiff was working. Both the plaintiff and the defendants were aware of the danger of rock slipping from the crane, but the practice nevertheless continued. Mr Smith was injured when he was hit by falling rock. He sued his employers for negligence and they raised the defence of *volenti*. The court rejected this defence stating, that although the plaintiff was aware of the risk and had continued to work whilst being aware of the risk, this did not constitute consenting to the risk of injury arising from his occupation.

As stated above, the courts now follow this reasoning and it is unlikely these days for the defence of volenti non fit injuria to succeed in employment cases.

One case where the employers raised the defence of volenti and the defence was successful was:

ICI v. Shatwell (1965)

In this case, two brothers, who were experienced shotfirers, ignored safety regulations and the instructions of their employers when testing detonators. The company procedure laid down for this operation was deliberately disregarded by the brothers and the result was that one of them was injured during an explosion. The court decided that:

1 the brothers were under no pressure from the company to adopt the procedure they had followed; and
2 they had failed to follow the correct and safe procedure; and that
3 in these circumstances, the defence of volenti raised by the company could succeed.

It follows from this, that departures from recognized procedures, in circumstances where employees have been trained to adopt the correct procedures, could result in an employee failing to recover compensation from his employer who may successfully raise the volenti defence. With reference to the remarks made earlier in this chapter concerning the use of written procedures as a possible aid to a defence against claims for negligence, this particular case is one instance of where that proved to be what happened. Although it also needs to be remembered that had the plaintiffs adhered to the procedures in question, then the harm which befell them would probably not have happened.

Driving cases   The statutory provisions of the Road Traffic Act 1988 s.149 apply to driving situations. Consider the following situation: the driver of a car is drunk, and the plaintiff accepts a lift, knowing the driver to be drunk, and is subsequently injured in an accident caused by the driver's negligence.The injured plaintiff sues the driver for negligent driving, in order to claim compensation for injuries sustained in the accident, and the driver raises the defence of volenti non fit injuria, claiming the passenger had consented to the risk of an accident occurring. The defence of volenti will not prevent the passenger from claiming compensation as the Road Traffic Act 1988 s.149 provides that volenti is not available where a passenger sues a driver in circumstances where insurance is compulsory.

Consider, however, a different story, that contained within *Morris* v. *Murray* (1990). In this case, the plaintiff and defendant had been drinking together when it was decided that they would go for a ride in a private plane. The defendant was piloting the plane when the plane crashed, injuring the plaintiff who sued for compensation. The defence of volenti succeeded as the drunkenness of the pilot was so obvious that a serious and obvious risk was created. The Road Traffic Act 1988 does not apply to aircraft.

In the case of *Pitts* v. *Hunt* (1990) the facts were as follows. The teenage defendant was driving a motor bike on which the plaintiff was a pillion passenger. The defendant had never passed a driving test and had no insurance cover. The defendant was driving dangerously, weaving in and out of the white lines down the centre of the road and driving too fast. The pillion passenger was urging the driver on. There was an accident in which the driver was killed and the passenger was seriously hurt. The passenger sued the estate of the driver. The defences relied on were *volenti non fit injuria* and *ex turpi causa non oritur actio*. The latter means that an action does not arise from a base cause (an illegal act). The defence of volenti was defeated by the statutory provisions of the RTA 1988. The defence of ex turpi causa non oritur actio succeeded, thereby preventing the defendant's estate from having to pay compensation to the plaintiff.

Contributory negligence

Contributory negligence is a far more successful defence than volenti non fit injuria. An example of the application of this defence would be where a plaintiff was suing a defendant, who recognized that he had been negligent, but also believed that the plaintiff had been negligent, and so raises the defence. This is further explained by the following case:

Owens v. *Brimmell* (1977)

In this case, the plaintiff and defendant had been drinking together and had visited several pubs in succession. The plaintiff had then accepted a lift home from the defendant, although he knew the defendant was drunk and should not have been driving. The plaintiff was then seriously injured in an accident. When the plaintiff sued the defendant for damages to compensate for his injuries, the court found that the defendant had driven negligently, but they also found that the injuries suffered by the plaintiff were partly his own fault. He had accepted a lift from the defendant, knowing he was drunk, and then he had failed to use his seat-belt. In instances such as this, the defendant must prove that the plaintiff had not taken reasonable care for his own safety and that this was a cause of his damage. Where the defendant is successful in establishing this, the court will then apportion fault between the parties. Then, the amount of compensation the defendant has to pay the plaintiff will be reduced accordingly, as was the case here.

The courts are far more likely to accept that a plaintiff was guilty of contributory negligence instead of being volens, because the plaintiff will not then be prevented from receiving some compensation. Volenti, if successful, is a complete defence to a negligence action. Therefore, a defendant who successfully pleads that the plaintiff was volens, will prevent the plaintiff from receiving any compensation at all.

Negligence: summation

As will be seen in later chapters, the subject of negligence is considerably more diverse than the legal statutes which follow. Whilst there are certain principles, which are common in various areas of English law, the statutes which are explained in subsequent chapters tend to be specific, in that they are aimed at either particular individuals or groups, businesses, services or products. The Consumer Protection Act 1987 for example concerns itself primarily with Product Liability, whilst the Health and Safety at Work Act 1974, as its name suggests, is concerned solely with health and safety issues. As a result, where specific statutes are concerned, management and quality practitioners can focus their attention on specific areas of concern and cater for these concerns in the way individual quality procedures are constructed.

The subject of negligence, however, is arguably considerably more diverse: 'the categories of negligence are never closed.' As a result this would seem to present management with a problem akin to that of the Gordian knot, or is it? Upon consideration, the prevention of negligence is similar to the implementation of a Quality Management System such as that required by the

British Standard BS EN ISO 9000 itself. Consider the latter for a moment. The standard is laid out in a systematic fashion, with specific elements which quality systems need to include. What the Standard does not set out to do is enforce uniformity, because it recognizes that individual quality assurance management systems will be as individual and diverse as the companies and businesses which seek to implement them. As a result, each company will need to create its own quality procedures to conform to the standard. The initial creation of such systems must have seemed to many Quality Managers to be a Gordian knot of its own.

Similarly, each company will need to review its procedures to ensure that the organization, its employees, and products or services conform to legal requirements. The need will not be to review specific legislation in relation to the company's own product or service, but rather to take an all encompassing view to try to ensure that its working practices and the actions of its employees cannot be construed as negligent. The penalties for failing in this area can be extremely costly. However, consider the effect of a negligent act from which the customer suffers harm. The penalties could well be incalculable – not only the immediate penalty of paying damages and legal costs, but also the damage to reputation and subsequent market share, and the possibility that negligence which causes physical injury could well put a company in breach of other legislation which, dependent upon the statutes involved, could itself lead to criminal prosecution with its own set of penalties.

So in creating a quality assurance management system which is mindful of the repercussions involved with negligence, where is the starting point? The answer to this question must be the same as to the question of where does quality begin? The answer must be with managerial commitment and the company's ethos with regard to the achievement of quality. In addition an understanding of what constitutes negligence and how that has been adjudged in the past is essential. Whilst third party liability insurance may ameliorate the financial effects of litigation it is only by creating and implementing good quality procedures, based upon best practice, in all areas, which can give a company some assurance that they will not fall foul of the laws concerning negligence.

Limitation Periods

When a person wishes to sue another party for compensation, when damage has been caused, there is only the right to do so for a certain period of time. A person will not be liable for an act of negligence for ever. Under the common law, there are no time limits in which a claim can be brought. This would be very onerous on a person who had been negligent. Parliament intervened and time limits were introduced. Under the Limitation Act 1980 s.2, tort actions must be brought within six years of the accrual of the cause of action. In cases which are dependent upon damage having been caused, the action accrues when the damage occurs. In some torts there is no need for damage to occur before being able to sue. Trespass is an example of this. In cases of trespass, a person is entitled to sue the trespasser even though the trespasser caused no damage to the plaintiff's property. In this instance, the action accrues when the trespass occurs.

Special rules of limitation apply in cases of personal injuries and death, latent damage, concealed fraud, and persons under a disability. In personal injury actions, the limitation period is normally three years from the date the action accrues or the date of the plaintiff's knowledge, whichever is later (Limitation Act 1980 s.11(4)). The inclusion here, of the plaintiff's 'knowledge', is very important, as the plaintiff may not develop symptoms immediately the action accrues. If the symptoms did not appear until years later (which is often the case when drugs are claimed to have caused harm to patients), then these potential plaintiffs would either have less time in which to bring the claim, or may have run out of time in which to bring an action. Thus, this section of the Limitation Act 1980 allows time to run from the date the plaintiff is aware of the disease or illness, and therefore will not stop them, from being able to bring an action for compensation.

Under the Limitation Act 1980 s.33, the court also has the right to allow a person to bring an action outside the three year limitation period if they consider it equitable to do so, having regard to all the circumstances of the case.

The only other special case to be referred to here is that concerning persons under a disability. The phrase, 'person under a disability', refers to 'infants' and persons of 'unsound mind'. Infants are classified as all those under the age of eighteen, and a person is of unsound mind if, by reason of mental disorder within the meaning of the Mental Health Act 1983, he is incapable of managing and administering his property and affairs (s.38(3)). Where a cause of action accrues to a person who is under a disability at the time the action accrues, then time will not start to run until that person ceases to be under a disability, or dies, whichever occurs first (s.28(1)).

A cause of action that had accrued to a person, who then died, was extinguished by the person's death. Also, there was no common law right of action granted to another party against the person who caused the death. This meant that the deceased's dependants were unable to recover compensation from the defendant, for the loss of support which they would have received from the deceased. Parliament introduced two statutes to address the unfairness of the common law rules.

The Law Reform (Miscellaneous Provisions) Act 1934, allows all existing causes of action (except defamation) to remain against a deceased defendant. This means that in instances where a defendant has caused a plaintiff harm, but the defendant then dies, the plaintiff will have a continuing cause of action against the estate of the deceased.

Where a plaintiff has a cause of action against a defendant, and then the plaintiff dies, the estate of the plaintiff will have a continuing cause of action against the defendant.

4

The Consumer Protection Act 1987 and the General Product Regulations 1994

Outline

The Consumer Protection Act 1987 was introduced into the United Kingdom as a result of a European Community directive which had been adopted by the Council of the EC for the purpose of harmonizing the laws on product liability throughout the member states. There was already pressure on the government of the United Kingdom to reform the existing law relating to defective products which, at that time, was covered by the tort of negligence, in cases where the claimant had no contractual remedy. The Law Commission proposed changes to the law first in a paper entitled 'Liability for Defective Products'. This was followed by a Royal Commission, chaired by Lord Pearson, which considered Civil Liability and Compensation for Personal Injury and included liability for defective products being strict liability instead of fault liability.

The thalidomide tragedy had invoked calls for a change in the way compensation was awarded for injuries sustained as a result of defective products being put on the market. Actions in negligence were unsatisfactory in many cases due to the necessity of proving fault, which was particularly difficult in actions against pharmaceutical companies where it was alleged that defects in prescribed drugs had caused harm to patients. The difficulty in such cases arose when trying to identify at which stage of the production process negligence had occurred which rendered the drug defective.

The adoption of the EC directive meant that under the United Kingdom's treaty obligations it was necessary for the government to implement a national law to accord with the requirements of the directive. Basically legislation was required in order to make producers of defective goods liable for personal injuries or damage to other property, which had been caused by the defective product, without the necessity of having to prove who was at fault. Providing the consumer could show that the product was defective, and had caused the harm complained of, then the producer would be liable. To further safeguard the consumer, the Act prohibits the inclusion of contractual clauses which attempt to limit or exclude liability to the consumer. The inclusion of such clauses may also be contrary to the Unfair Contract Terms Act 1977.

The Consumer Protection Act 1987 needs to be examined in greater detail before considering the effects of contravening the Act and how such contraventions may possibly be reduced or eliminated by the implementation of a quality assurance management system (QAMS) such as BS EN ISO 9001:1994. Further, it needs to be made absolutely clear from the outset that there are no guarantees that by devising and following a quality assurance management system, civil or criminal liabilities will not be incurred. Although the adoption of, and compliance with, recognized national and international standards may well prove to be beneficial when assessing the safety of a product with regard to the General Product Safety Regulations 1994 and may provide a defence under s.4 Consumer Protection Act 1987. What a well devised QAMS can do, with regard to consumer protection, is:

1 raise the level of awareness of obligation with regard to the consumer, and as a result promote the development of procedures which in turn will minimize the risks of litigation and promote a safer system of general working practices;
2 by means of its documented systems help to provide the necessary evidence to mount a successful defence against civil claims or criminal prosecutions;
3 provide documentary evidence to be used in civil actions against those who may be the actual source of blame in any justifiable claim made by a consumer.

Strict Liability

The one major advantage the Consumer Protection Act 1987 has when compared to an action brought on the grounds of negligence is that a claimant does not have to prove the manufacturer was negligent before being awarded compensation. This form of liability is known as *strict liability*, or *liability without fault*. The mere fact that the goods are faulty is sufficient to render the seller liable in contract law. There is no need to show that the seller is at fault in supplying the goods as explained in:

Frost v. *Aylesbury Dairy Co.* (1905)

Mr Frost's wife died as a result of drinking milk infected with typhoid germs. Although it was established that the dairy was extremely careful during their production process, the court held the dairy liable because under the contract of sale there was an implied condition that the milk was reasonably fit for human consumption. As the milk was not fit to be consumed the dairy was in breach of their contract of sale.

Even in situations where everything conceivable has been carried out to ensure that a defective product is not allowed on the market, businesses can still be liable under the principle of strict liability. This is why it is so important to minimize the risks of this happening. The opposite of 'strict liability' is 'fault based liability' which is where, in order for a complainant to succeed in obtaining compensation for harm or injury suffered, the complainant must prove the defendant was to blame for the injury received. This can be seen in the next case.

Donoghue v. *Stevenson* (1932)

Mrs Donoghue suffered gastro-enteritis as a result of drinking ginger beer in which she had found the remains of a decomposed snail. Her friend had bought the ginger beer for her at a cafe. Mrs Donoghue's sued the manufacture in an action in which she claimed the manufacturer had been negligent during the manufacturing process by allowing the snail into the bottle. This was the first time that the court had decided that a manufacturer could be liable to a consumer suffering personal injuries as a result of a defective product.

Until the case of *Donoghue* v. *Stevenson* came before the courts the only possible remedy would have been in contract law. Mrs Donoghue would not have had to show anything other than the existence of a contract. There would have been no need to prove that the manufacturer had been negligent because the mere fact that there was a contract in existence rendered the supplier of the ginger beer liable under the Sale of Goods Act 1893. The problem in Mrs Donoghue's case was that she did not have a contract with anyone. Her friend had a contract with the cafe owner but because she had not suffered any injury, she had no cause of action. An action in negligence was Mrs Donoghue's only chance of being awarded compensation. The person who would be liable under a negligence claim would not be the cafe owner as he had not been at fault. He was not to blame for the snail getting into the ginger beer, therefore, the manufacturer was sued.

If the case of *Frost* v. *Aylesbury Dairy Co.* (1905) had been pursued as a claim in negligence, Mr Frost would not have succeeded. The dairy had taken all reasonable care and had not been negligent. Under contract law, the dairy were liable as they had sold the milk in the course of a business and were in breach of the implied terms relating to merchantable quality and fitness for the purpose. However, under the Consumer Protection Act 1987 all that would need to be established is that the product was 'defective' and had 'caused the consumer harm'.

The Consumer Protection Act 1987 – Part 1

The Consumer Protection Act provides for both civil claims and criminal prosecutions. Part 1 of the Consumer Protection Act 1987, which covers civil claims, put into effect the EC Directive on Product Liability, and came into force on 1 March 1988. The aim of this part of the Act is to make the producer liable for personal injury or damage to other property caused by a defective product without the consumer having to prove any fault on behalf of the producer. It is important to note that under this Act the consumer is not solely the person to whom the product has been supplied. The Act covers injury which is caused to any person, or damage which is caused to any property, other than the product itself, irrespective of whether or not that person was the actual purchaser of the product.

The following sets out to explain some of the most important sections of the Consumer Protection Act 1987 in order that:

1 persons who may be liable under the Act can be identified;
2 a 'defective product' can be defined;
3 damage can be defined;
4 liability can be considered.

Persons Who May Be Liable under the Act

Figure 4.1 shows the main groups and subgroups of those persons who may be liable under the Consumer Protection Act 1987.

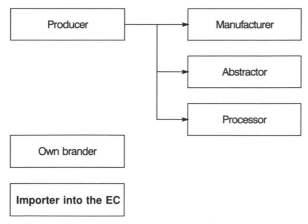

Figure 4.1 Persons having liability

The producer, for the purposes of the Act, is defined in s.1(2) where it states that the 'producer', in relation to a product, means

1 the person who manufactured it;
2 in the case of a substance which has not been manufactured but has been won or abstracted, the person who won or abstracted it;
3 in the case of a product which has not been manufactured, won or abstracted but essential characteristics of which are attributable to an industrial process having been carried out (for example, in relation to agricultural produce), the person who carried out that process.

'Producer' may be further explained as follows:

1 The *manufacturer* includes the final manufacturer of the product and also any manufacturer of a faulty component part which has caused the finished product to be defective. The assembler may also be liable under this section. Section 2(5) states that their liability is joint and several, which means that the claimant could take action against any one of the parties for the full amount sought. Liability between the parties themselves may then be apportioned under the Civil Liability (Contribution) Act 1978.
2 The *person who won or abstracted* a substance covers those who abstract raw materials and minerals such as oil or coal.
3 The *processor* covers those not covered by any of the above. For example, a 'picker' of fresh peas would not be covered by the section, but a 'canner' would be subject to the Act because tinned peas have undergone an industrial process.

Section 2(2) – who is liable under the Act?

Those described under s.2(2) include:

(a) the producer of the product;

(b) any person who, by putting his name on the product or using a trade mark or any other distinguishing mark in relation to the product, has held himself out to be the producer of the product;

(c) any person who has imported the product into a member state from a place outside the member states in order, in the course of a business of his, to supply it to another.

Section 2(2) explained

The 'producer' in (a) above includes all persons referred to in the definition section.

The person who 'holds himself out' as being the producer of a product in (b) above could include a supermarket who has put their name or brand onto the product in question so that their customers believe they are the producers of that product. This subsection would appear to make all supermarkets and other stores liable for damage caused by any defects in their own-branded products. However this is not the situation where the label clearly states that the product has been made for them by another party. Section 2(2)(b) may not then apply as they will not be 'holding themselves out' as being the producer of the goods. This demonstrates the importance of labelling and packaging all products clearly which may then help prevent civil liability from arising.

Section (c) applies to an importer into the EC, importing for business purposes. This aids consumers by identifying a potential defendant within the EC who will be liable to the consumer, thus preventing the consumer from having to take legal action against a foreign producer with all the difficulties which that entails.

As can be seen from the above, liability for damage or injury caused by defective products is only being placed on 'retailers' who have put their own mark or brand on the product in question. However, the next section is placing liability on the supplier of defective goods in certain circumstances. The supplier in this section may well include the retailer.

Section 2 (3)

Section 2(3) provides that:

> Where any damage is caused wholly or partly by a defect in a product, any person who supplied the product (whether to the person who suffered

the damage, to the producer of any product in which the product in question is comprised or to any other person) shall be liable for the damage if:

1 the person who suffered the damage requests the supplier to identify one or more of the persons (whether still in existence or not) to whom subsection (2) above applies in relation to the product;
2 that request is made within a reasonable period after the damage occurs and at a time when it is not reasonably practicable for the person making the request to identify all those persons; and
3 the supplier fails, within a reasonable period after receiving the request, either to comply with the request or to identify the person who supplied the product to him.

A prerequisite of any QAMS is the keeping of accurate records. Indeed as most quality practitioners will be aware, ineffective document control is a major source of non compliance during second or third party audits. The possibility of litigation adds an additional dimension to the need for both accurately prepared and carefully controlled documents. One of the aims of the Consumer Protection Act 1987 was to prevent defective products from reaching the market place. It was hoped that by making the supplier of goods (which would normally be the retailer) liable in situations where the producer could not be identified, that this would encourage suppliers to deal only with reputable parties.

A notable exception which is made within the Act concerns the supplier of agricultural produce. Section 2(4) of the Act says 'neither subsection (2) nor subsection (3) above shall apply to a person in respect of any defect in any game or agricultural produce if the only supply of the game or produce by that person to another was at a time when it had not undergone an industrial process'.

Therefore the suppliers of BSE contaminated beef, or the suppliers of eggs contaminated with salmonella would not be liable under the Consumer Protection Act 1987 where these products had not undergone an industrial process. This does not mean that the suppliers would not be liable at all for supplying defective products but what is does mean is that a person harmed by a defective product would have to seek redress in other ways. For example:

1 a consumer who had a contract with the supplier would have an action for breach of the implied terms of 'satisfactory quality' and 'fit for the purpose' under the Sale of Goods Act 1979 (as amended by the Sale and Supply of Goods Act 1994);

2 if the person suffering harm, caused by the defective product, was not the person with whom the contract had been made, then any redress would need to be sought under the tort of negligence.

The Consumer Protection Act 1987 specifically recognizes these other causes of action in section 2(6) which says that liability may also arise outside the jurisdiction of the Act.

It needs to be reiterated that as liability arises from more than one cause of action, it is extremely important that management and quality practitioners are aware of all possible sources from which liability may be incurred in order that the most effective quality procedures and working practices can be devised and operated.

The Definition of a Defective Product

The Consumer Protection Act 1987 provides a number of definitions in order to clarify what is meant by the term defective product. The definitions include the following:

1 product;
2 goods;
3 defect;
4 safety;
5 damage.

Product

For the purposes of the Act a product is defined in section 1(2) which provides that: 'product' means any goods or electricity and (subject to subsection (3) below) includes a product which is comprised in another product, whether by virtue of being a component part or raw material or otherwise.

Goods

These are defined within section 45(1), which is the interpretation section of the Act. 'Goods' includes substances, growing crops and things comprised in land by virtue of being attached to it and any ship, aircraft, or vehicle.

Human tissue or blood are not specified in the Act so it is unlikely that they would be included, even under substances. The Pearson Committee recommended that products should include human tissue and blood so that the persons distributing the said blood and tissue would be liable for defects contained therein as if they were products. Considering some of the problems that have arisen involving blood transfusions, governmental reluctance to allow claims under the Act for harm caused by defective blood products is easy to understand. A fear of such claims opening the floodgates to future actions could have affected the government's decision to leave such claims to be resolved through the tort of negligence.

Defect

According to section 3(1) of the Act, there is a defect in a product for the purposes of this Part if the safety of the product is not such as persons generally are entitled to expect; and for those purposes 'safety', in relation to a product, includes safety with respect to products comprised in that product and safety in the context of risks of damage to property, as well as in the context of risks of death or personal injury.

Safety

Section 3(2) states that:

> In determining for the purposes of subsection (1) above what persons generally are entitled to expect in relation to a product, all the circumstances are to be taken into account, including:
>
> 1 the manner in which, and the purposes for which, the product has been marketed, its get-up, the use of any mark in relation to the product and any instructions for, or warnings with respect to, doing or refraining from doing anything with or in relation to the product;
> 2 what might reasonably be expected to be done with or in relation to the product; and
> 3 the time when the product was supplied by its producer to another;
>
> and nothing in this section shall require a defect to be inferred from the fact alone that the safety of a product which is supplied after that time is greater than the safety of the product in question.

Section 3(2) above makes it clear that the safety of the product itself may depend on how the product was marketed and may include such issues as: What use(s) was the product intended for and by whom? Was it intended for use by the general public, and thus being described as simple to use, or was it described as being suitable for the more experienced DIY enthusiast? The instructions must be accurate and capable of being clearly understood. If the instructions are ambiguous and a consumer was harmed by the product after following the instructions, then a court could decide that the product was not as safe as persons are generally entitled to expect and the producer would be liable under the Act.

Section 3(2)(b) covers situations where a product may not have been used as intended by the producer. The product must be used in a way in which such a product might reasonably be expected to be used. Therefore, if a product is put to an unusual use and the court decides that the product had not been used in a reasonable way, the court may determine that the product was not defective for the purposes of the Act and therefore no liability would arise for any harm suffered.

Improvements are continually being made to products, and so, by the very natutre of technology, the latest model of a particular product released on to the market will be the safest. Section 3(2)(c) covers this type of situation. The time when a product was supplied by the producer will be taken into account when determining whether that product is as safe as people generally are entitled to expect. The fact that a later, safer model has since been designed and is available in the market place does not mean that the product in question is defective. There must be other evidence to support the finding that the product is defective.

The effects of section 3 of the Act will now be examined. Subsection (a) is referring to the packaging, labelling and marketing of the product in question. Care must be taken to ensure that no exaggerated claims are made concerning the capabilities or use of the product, and that any instructions are clear and comprehensive enough to enable the product to be used safely. It is worth noting that the Trade Descriptions Act 1968 is also relevant in the area of exaggerated claims. It would be easy to dismiss such claims as being the province of marketing and sales. However, with regard to the Consumer Protection Act 1987, contract review procedures should initially ensure that businesses can provide a safe product which meets customers' requirements. Quality procedures should not only be in place where the labelling of goods and the writing of instructions for those goods are concerned, but should also exist in areas of:

1 design; and
2 manufacture and inspection; or
3 purchase and the verification of purchased products.

Such procedures should allow for the transfer of the correct information to ensure that any label or instructions accurately represent the product's capabilities.

The requirements of subsection 3(1)(b) of the Act may prevent claims being made by consumers who have put their products to strange or unusual uses and then hope to be awarded sometimes huge amounts in compensation. Tales of owners putting their cats in the microwave in order to dry them and then seeking damages from the manufacturer for failing to warn them of the dangers of placing animals in the microwave, make for entertaining reading in daily newspapers, but are of concern to producers facing litigation. Businesses would be advised to ensure explicit instructions for use were included on labels and packaging, but it is suggested that this may be one area in which it would be impossible to warn against all eventualities. However, the requirement under the Consumer Protection Act 1987, that 'the use is one to which the product may reasonably be expected to be put', should prevent such ludicrous claims being made under the Act. The final requirement contained within section 3(2) is that the time at which the product was supplied by the producer must be taken into account in order to help determine whether the product is 'safe' for the purposes of the Act. This is because the safety of the product must be judged by the standard of safety which prevailed at the time the product was put onto the market. This prevents producers from being liable for products which may be considered defective if judged by current standards, which in turn may take into account safety features unknown at the time the original product was produced.

Damage

This is defined in section 5(1) of the Consumer Protection Act which provides:

1 Subject to the following provisions of this section, in this Part 'damage' means death or personal injury or any loss of or damage to any property (including land).
2 A person shall not be liable under section 2 above in respect of any defect in a product for the loss of or damage to the product itself or for loss of any damage to the whole or any part of any product which has been supplied with the product in question comprised in it.
3 A person shall not be liable under section 2 above for any loss of or any damage to any property which, at the time it is lost or damaged, is not:
   (a) of a description of property ordinarily intended for private use, occupation or consumption; and
   (b) intended by the person suffering the loss or damage mainly for his own private use, occupation or consumption.

4 No damages shall be awarded to any person by virtue of this Part in respect of any loss of or damage to any property if the amount which would fall to be so awarded to that person, apart from this subsection and any liability for interest, does not exceed £275.

Section 5 is therefore setting out the possible types of damage that may be claimed for under the Act, and the limitation to possible claims.

Explanation of section 5

The Act imposes no financial limit on the amount which could be claimed for injuries or death suffered as a result of a defective product, but does impose a minimum amount of damage to have been caused to other property before being able to claim under the Act. Also note that subsection (2) states that it is only damage to other property or products which is covered and not damage to the product itself that the producer will be liable for. The effect of this can be seen by the following examples.

If a new car was purchased from a garage and was involved in an accident due to a faulty tyre then neither the damage to the car itself nor the damage to the tyre could be recovered under the provisions of the Consumer Protection Act because the tyre had been supplied with the car. In this situation compensation would be limited to any personal injuries sustained in the accident and to any damage to other vehicles involved in the accident. Contrast this situation with the next scenario in which a car has just been fitted with replacement tyres at a garage and was involved in an identical accident to the above. Compensation could now be recovered from the producer of the defective tyre, not for damage to the tyre itself but for the damage caused to the car, as in this case the car had not been supplied with the tyre. In addition a claim could be made for personal injuries and damage to other vehicles as above. In both cases, compensation for the damage to the product itself would need to be taken against the retailer under contract law, where there was a contract in existence and in an action for negligence against the manufacturer where no contractual remedy was available.

Exclusions of liability

Section 5(3) excludes liability for damaged products intended for commercial use. It is not always easy to identify those products which are intended 'mainly for . . . private use' as required by section 3(b). This will depend on the particular circumstances of the case in question. Section 5(4) excludes claims where damage to other property is worth less than £275, the purpose of a minimum amount being to prevent a plethora of small claims being made.

Defences

What defences to an action brought under the Consumer Protection Act 1987 are available?

Section 7 provides the following: 'The liability of a person by virtue of this Part to a person who has suffered damage caused wholly or partly by a defect in a product, or to a dependant or relative of such a person, shall not be limited or excluded by any contract term, by any notice or by any other provision. when that person has suffered damage which has been caused wholly or in part by a defect in a product.'

As a result there is no possibility of excluding liability and then claiming this as a defence to any action. The defences which are available are covered by section 4 of the Act and are listed, together with explanations, below.

Defences under section 4

(1) In any civil proceedings by virtue of this Part [of the Act] against any person ('the person proceeded against') in respect of a defect in a product it shall be a defence for him to show:

(a) that the defect is attributable to compliance with any requirement imposed by or under any enactment or with any Community obligation; or

(b) that the person proceeded against did not at any time supply a product to another; or

(c) that the following conditions are satisfied, that is to say:
   (i) that the only supply of the product to another by the person proceeded against was otherwise than in the course of a business of that person's; and
   (ii) that section 2(2) [of the Act] does not apply to that person or applies to him by virtue only of things done otherwise than with a view to profit; or

(d) that the defect did not exist in the product at the relevant time; or

(e) that the state of scientific and technical knowledge at the relevant time was not such that a producer of products of the same description as the product in question might be expected to have discovered the defect if it had existed in his products while they were under his control; or

(f) that the defect:
    (i) constituted a defect in a product ('the subsequent product') in which the product in question had been comprised; and
    (ii) was wholly attributable to the design of the subsequent product or to compliance by the producer of the product in question with instructions given by the producer of the subsequent product.

Thus, it would be a defence in (a) for the producer if he could show that the defect in the product was caused by complying with a statute or EC Regulation or in (b) if he could show that he had not supplied the product in question, but, for example if the product had been stolen. The defence in subsection (c) would apply where the producer had given the product as a gift or had sold home-made produce at a fund raising event and had not supplied the product in the course of a business. Subsection (d) provides a defence for the producer where the defect did not exist at the 'relevant time'. The 'relevant time' is the time the product was supplied or put into circulation by the producer, own-brander or importer under section 2(2). Where the person proceeded against (the defendant) is a supplier, the relevant time is when the product was last supplied not by the supplier but by the producer, own brander or importer as above. Also, it is a defence if the defect in the product is a result of interference by a third party, the defect in the product has been caused by misuse, or where the defect has been caused by fair wear and tear. Where goods are perishable and have passed through several different parties in the distribution chain before reaching the consumer, they may have left one party in perfectly good condition but are less than perfect by the time they reach the consumer. If the producer can show that the goods were in good condition when they left his control then he could escape liability using this defence. In a situation where a consumer has shown that the product was defective, in accordance with the requirements of the Act and damage has been caused, the producer must then prove the defence according to the civil burden of proof, namely, the balance of probabilities. It is therefore essential that all 'producers' under the Consumer Protection Act 1987 keep up to date, detailed records of products which they receive and despatch. They will then have the necessary documentation to provide evidence, if required, of the date on which the goods were received and the condition of the goods when inspected and also the date on which the goods were forwarded and their condition at that time. These procedures should form part of every quality assurance system in situations such as this.

Subsection (e) of the Act is probably the most controversial. The defence under subsection (e) is known as the 'development risks' defence. The EC member states were given a choice under the EC Directive on whether to incorporate this defence into their own legislation or not. Britain decided to include the defence as it was feared that the development of new products would be jeopardized if producers were made liable for harm suffered or damage caused to property where they had taken every possible precaution but nevertheless injury or damage had resulted. The purpose of the EC Directive was to enable consumers to claim compensation for injuries or damage caused by defective products without the necessity of proving fault, or in other words, making the producers strictly liable. It was argued that producers were in the best position to make compensatory awards as they could insure themselves against any possible claims. The cost of the insurance premiums would be incorporated into their overall production costs which could then be passed on to the consumers. The effect of incorporating the 'development risks' defence into the Consumer Protection Act 1987 was to limit the application of strict liability. Under the Act, if a producer is unaware of there being a defect in the product, because of the state of scientific or technical knowledge at the time the product is put onto the market, then the producer will escape liability. As stated at the beginning of this chapter, the introduction of strict liability for defective products was seen to be desirable after the thalidomide tragedy where victims of the drug were left without an effective means of obtaining compensation as they could not prove the drug company was at fault when manufacturing the drug. Unhappily, if a similar situation arose today, the victims would probably be in the same position due to the incorporation of the 'development risks' defence.

The defence is similar to that of 'the state of the art defence' in negligence actions. This can be demonstrated by the case of:

Roe v. Ministry of Health (1947)

Mr Roe was admitted to hospital for a minor operation. He was paralysed from the waist down when an anaesthetist administered anaesthetic which had been contaminated by phenol, a disinfectant. The anaesthetic had been stored in glass ampoules which were kept in phenol until required. The phenol had seeped through invisible cracks in the glass contaminating the anaesthetic. No negligence was found as it was not known amongst the profession that there was a risk of this happening until four years later in 1951.

Contributory negligence

An additional defence would be where the defendant could establish that there was contributory negligence on the part of the claimant. Indeed section 6 of the Act specifically includes the provision for such a defence.

Section 6(4) This section states: Where any damage is caused partly by a defect in a product and partly by the fault of the person suffering the damage, the Law Reform (Contributory Negligence) Act 1945 and section 5 of the Fatal Accidents Act 1976 (contributory negligence) shall have effect as if the defect were the fault of every person liable by virtue of this Part for the damage caused by the defect.

The reference to 'every person' includes the person who has been 'damaged'.

This section shows that the Act creates 'strict liability' and not 'absolute liability', which allows for some defences to operate. However, contributory negligence is not capable of being a complete defence, or, in other words, allowing the defendant to argue that the claimant or plaintiff was 100% to blame, so enabling a 100% reduction in any compensation that would have been awarded. Thus the claimant will always receive some compensation.

Limitation of Actions

A person who has suffered personal injury or damage to private property does not have an indefinite time period in which to sue. The Limitation Act 1980 imposes time limits on the claims of potential plaintiffs. Thus if a plaintiff has failed to issue proceedings within a certain time limit, the defendant can then claim that the plaintiff is time barred from taking action. Injustices occurred when an earlier Limitation Act was applied to situations where the symptoms of the injury materialized only after many years. An example of this can be seen in the case of:

Cartledge v. *E Jopling & Sons Ltd* (1963)

The defendant's breach of duty caused one of their servants, Mr Cartledge, to inhale harmful dust which caused the onset of pneumoconiosis. The symptoms of the disease only became apparent after the time limit for taking actions had elapsed and the House of Lords was forced to hold that the plaintiff's action was time barred by statute (the Limitation Act). Later in that year a new Act of Parliament was created in order to prevent further injustices of this nature.

As part of the Consumer Protection Act 1987, schedule 1 stipulates that actions with regard to personal injuries would be covered by section 11 of the Limitation Act 1980. Briefly, the section applies to any action for damages for:

1 negligence;
2 nuisance;
3 breach of duty.

Section 11A of the Limitation Act makes special provision for actions in respect of defective products, providing that this section will apply to actions for damages under the Consumer Protection Act 1987.

The Limitation Act 1980 Section 11A(4)   Subject to subsection (5) below, an action to which this section applies in which the damages claimed by the plaintiff consist of or include damages in respect of personal injuries to the plaintiff or any other person or loss of or damage to any property, shall not be brought after the expiration of the period of three years from whichever is the later of:

(a) the date on which the cause of action accrued; and

(b) the date of knowledge of the injured person or, in the case of loss or damage to property, the date of knowledge of the plaintiff or (if earlier) of any person in whom his cause of action was previously invested.

The date on which the cause of action accrued in (a) means the date on which the damage occurred, so the claimant has three years, from that date, in which to start an action for damages. The date of knowledge of the injured person in (b) is defined to include:

1 the fact that the damage was significant; and
2 the fact that it was caused by the defect; and
3 the identity of the defendant.

Care needs to be taken with regard to the calculations as time is not prevented from running by the plaintiff's ignorance that as a matter of law the product was defective.

Section 11(5)   If in a case where the damages claimed by the plaintiff consist of or include damages in respect of personal injuries to the plaintiff or any other person and the injured person died before the expiration of the period mentioned in subsection (4) [of the Limitation Act] above, that subsection shall have effect as respects the cause of action surviving for the benefit of his estate by virtue of section 1 of the Law Reform (Miscellaneous Provisions) Act 1934 as if for the reference to that period there were substituted a reference to the period of three years from whichever is the later of:

(a) the date of death; and

(b) the date of the personal representative's knowledge.

So now, a plaintiff has three years from the time when the injury occurs, or when the symptoms of the injury become apparent, in which to commence a claim against the defendant. However, all of this is subject to the provision of the following section.

Section 11A(3)   An action to which this section applies shall not be brought after the expiration of ten years from the relevant time, within the meaning of section 4 of the said Act of 1987; and this subsection shall operate to extinguish a right of action and shall do so whether or not that right of action had accrued, or time under the following provisions of this Act Section 11A(4&5) had begun to run, at the end of the said period of ten years.

The effect of this section is to create an overall long stop which prevents a plaintiff from commencing an action where a product has been in circulation for a period of in excess of ten years. This long stop provision is an absolute bar to an action under the Consumer Protection Act 1987 and cannot be overruled by the discretion of the court. In other cases of personal injury actions, section 33 of the Limitation Act 1980 gives a court the discretion to exclude the time limits for actions in respect of personal injuries or death if the court considers that to apply the limitations would unfairly prejudice the plaintiff or any other person he may represent. Guidelines are given in section 33(3) where the court is considering using its discretion to exclude the limitation periods. These will include:

(a) the reason and length of the delay on the part of the plaintiff;
(b) the extent to which the cogency of the evidence adduced by either the plaintiff or the defendant may be affected by the delay;
(c) the conduct of the defendant after the cause of action arose, including the extent (if any) to which he responded to requests reasonably made by the plaintiff for information or inspection for the purpose of ascertaining facts which were or might be relevant to the plaintiff's cause of action against the defendant;
(d) the duration of any disability of the plaintiff arising after the date of the accrual of the cause of action;
(e) the extent to which the plaintiff acted promptly and reasonably once he knew whether or not the act or omission of the defendant, to which the injury was attributable, might be capable at that time of giving rise to an action for damages;
(f) the steps, if any, taken by the plaintiff to obtain medical advice and the nature of any such advice he may have received.

Section 5   Section 5 of the Limitation Act lays down the limitation period for contractual claims. An action founded on simple contract shall not be brought after the expiration of six years from the date on which the cause of action accrued. Thus, a party to a contract has six years from the date when the contract was breached in which to commence an action against the other party to the contract. From this it can be seen that a person will not remain liable forever for their breach of contract or the results of their negligent acts. Once again it can be seen how important it is to keep complete and accurate records. If a producer, own brander, or importer into the EC supplies a defective product to another and can be liable for up to ten years after doing so, if accurate records are kept it may be shown that in fact the ten year period had expired several months previously, or records may show that the product was not defective whilst in the defendant's control. By devising a quality assurance system and ensuring its operation when dealing with the importation, storing, selling or purchasing of goods, evidence is being gathered, which in the event of any dispute arising over alleged defective products, could be produced in court to avoid liability.

Section 1(3) also provides for an important limitation by stating: 'a person who supplies any product in which products are comprised, whether by virtue of being component parts or raw materials or otherwise, shall not be treated by reason only of his supply of that product as supplying any of the products so comprised'.

Application of that section can be demonstrated by the following example. If (A) buys a new car, which unknown to him at the time, has a defective gear-box he would be able to hold both the car dealer (B) and the manufacturer of the defective gear-box liable for any injuries sustained as a result of the defect. If action was taken against (B) under the provisions of this section (B) would be able to escape liability by identifying the person who supplied him with the car. The identification of the manufacturer is not necessary. Again the importance of accurate record keeping cannot be over-emphasized. Persons who may be so affected should ensure records are kept of the suppliers of:

1 the finished products,
2 component parts, and
3 raw materials

and all dates of distribution or supply.

The Consumer Protection Act 1987 – Part 2

Part 1 of the Act is concerned with civil liability and gives the injured party the right to sue one or more possible defendants, ensuring that persons suffering from personal injuries or damage to private property caused by a defective product will be compensated for their loss. The aim of Part 2 of the Act in particular was to prevent defective goods from reaching the market place in the first place. Section 10 of the Act which was concerned with the General Safety Requirement (GSR) has now been disapplied by the General Product Safety Regulations 1994. The remaining provisions of the Consumer Protection Act 1987 regarding the power to make safety regulations and issue notices is unaffected.

For the most part it will be the safety requirement aspects of Part 2 which will give rise to the most immediate concern from the point of view of the quality of a product or service. As already stated, section 10 of the Consumer Protection Act 1987 was disapplied by the General Product Safety Regulations 1994, details of which can be found towards the end of this chapter.

The Consumer Protection Act 1987 – Part 3

Whilst Parts 1 and 2 of the Act concerned themselves with the associated subjects of defective products and safety requirements, Part 3 attempts to provide a measure of consumer protection from the point of view of misleading indications as to price. Indeed as a result of this part of the Act, section 11 of the Trade Descriptions Act 1968 is repealed.

Part 3 creates two criminal offences. The first, under section 20(1): '. . . that a person shall be guilty of an offence if, in the course of any business of his, he gives (by any means what so ever) to any consumer an indication which is misleading as to the price at which any goods, services, accommodation of facilities are available (whether generally or from persons in particular)'.

Notice the broad spectrum of businesses to which to which this offence is applicable. Also important to notice are the phrases '. . . by any means whatever . . .' and '. . . whether generally or from particular persons . . .' which refer to how misleading indications may be given. These would both suggest that anyone who has direct contact with, or whose comments or actions may directly influence the consumer, should be well trained in their respective product or service, together with the methods by which they present information to the consumer.

The second, covered by section 20(2) requires a combination of events, viz.:

(a) '. . . an indication to any consumer which, after it was given, has become misleading . . .'; and,

(b) '. . . some or all of those consumers might reasonably be expected to rely on the indication at a time after it has become misleading'; and

(c) the person responsible '. . . fails to take all such steps as are reasonable to prevent those consumers from relying on the indication.'

What actually constitutes a misleading statement is explained below. However, at this juncture a point which should be of grave concern to all in management generally, and to those involved in specifying procedures in particular, is that covered by section 20(3) of the Act, which provides:

'For the purposes of this section it shall be immaterial:
(a) whether the person who gives or gave the indication is or was acting on his own behalf or on behalf of another . . .'

This clearly indicates the need for comprehensive procedures with regard to pricing to be in place. In addition it highlights a legal necessity for the personnel who are operating those procedures, irrespective of their own standing in the company, to be equipped with the knowledge and expertise required for the procedures to be implemented correctly.

Meaning of Misleading

What actually constitutes a misleading statement with regard to price is covered by section 21 of the Act. Within this section it is stated that it is what '. . . consumers may reasonably infer from any indication or omission . . .' which may be adjudged to be misleading. Inferences specifically covered by the act are:
(a) that the price is less than in fact it is;
(b) that the applicability of the price does not depend on the facts or circumstances on which its applicability does in fact depend;
(c) that the price covers matters in respect of which an additional charge is in fact made;
(d) that a person who in fact has no such expectation:
  (i) expects the price to be increased or reduced (whether or not for a particular amount); or
  (ii) expects the price, or the price as increased or reduced, to be maintained (whether or not for a particular period);
(e) that the facts or circumstances by reference to which the consumer might reasonably be expected to judge the validity of any relevant comparison made or implied by the indication are not what in fact they are.

Penalties and Time Limits

Being a section of the Act which imposes criminal liability, conviction carries with it the prospects of a fine. The time limits for a prosecution to be brought are specified under section 20(5) as:

(a) the end of the period of three years beginning with the day on which the offence was committed; and

(b) the end of the period of one year beginning with the day on which the person bringing the prosecution discovered that the offence had been committed.

Once again there is a reason for ensuring that records with regard to pricing are both accurate and retained.

The General Product Safety Regulations 1994

The Council of Ministers adopted the EC Directive on General Product Safety in 1992 and the Regulations implement the provisions of that Directive. The Regulations themselves came into force on 3 October 1994. They impose criminal liability on new safety requirements for products which are placed on the market by suppliers or distributors, and intended to be used, or likely to be used by consumers. However, the Regulations only cover:

1 products which have been sold in the course of a commercial activity; and
2 products sold to persons not buying those products in the course of a commercial activity.

A commercial activity is defined as meaning any business or trade. Conversely, a product which was being utilized by a consumer, but which was ordinarily intended for use within a commercial environment would not be subject to the Regulations. A major advance in consumer protection was made by these Regulations in that they apply not only to new goods but also to second-hand and reconditioned merchandise. However, there are a number of exceptions specified by Regulation 3 which states that the Regulations do not apply to the following types of products:

1 second-hand products which are antiques;
2 products supplied for repair or reconditioning before use, provided the supplier informs the customer of this; (exempting these products from the provisions of the Regulations could be achieved by making it clear to the customer by ensuring the information is contained in the contract of sale);
3 products that are subject to specific provisions of EC law which covers all aspects of their safety;
4 products that are subject to specific provisions of EC law which cover an aspect of safety.

Safe Products

The Regulations impose a statutory duty on the producer to ensure that all products placed on the market are safe. The Regulations make it an offence to fail to comply with the general safety requirement and producers and distributors can be guilty of an offence by offering to supply, or agreeing to place, a dangerous product on the market, or even possessing a dangerous product for supplying or placing on the market.

Regulation 2 defines what is meant by 'safe' by stating the following:

'. . . any product which, under normal or reasonably foreseeable conditions of use, including duration, does not present any risk or only the minimum risks compatible with the product's use, considered as acceptable and consistent with a high level of protection for the safety and health of persons, taking into account in particular:

(a) the characteristics of the product, including its composition, packaging, instructions for assembly and maintenance;

(b) the effect on other products, where it is reasonably foreseeable that it will be used with other products;

(c) the presentation of the product, the labelling, any instructions for its use and disposal and any other indication or information provided by the producer; and

(d) the categories of consumers at serious risk when using the product, in particular children, . . . and the fact that higher levels of safety may be obtained or other products presenting a lesser degree of risk may be available shall not of itself cause the product to be considered other than a safe product.

The above is very similar to the provisions set out in section 10 of the Consumer Protection Act 1987 but is slightly more specific and focuses on consumers who are considered to be most at risk from possible harm. More everyday products – such as cleaning materials – may now be covered by these Regulations, a necessity given that children may come into contact with substances such as cleaning materials around the home.'

The Producer

Regulation 2 also defines the 'producer' as:

(a) the manufacturer of the product, when he is established in the [European] Community, and includes any person presenting himself as the manufacturer by affixing to the product his name, trade mark or other distinctive mark, or the person who reconditions the product;

(b) when the manufacturer is not established in the Community:
   (i) if the manufacturer does not have a representative established in the Community, the importer of the product;
   (ii) in all other cases, the manufacturer's representative; and
(c) other professionals in the supply chain, insofar as their activities may affect the safety properties of a product placed on the market.

Penalties

Again, it can be seen that the 'producer' under the General Product Safety Regulations is defined in similar terms to the 'producer' under the Consumer Protection Act, although the 'producer' does not appear to be as strictly defined under the Regulations, as liability under (c) above, depends on the activities of the persons in the supply chain affecting the safety of a product. No such requirement is specified in the Consumer Protection Act. It needs to be borne in mind that, under the Regulations, it is *criminal* liability that is being imposed and a person who is found guilty of breaching the Regulations can be sentenced by the Magistrates Court to three months imprisonment, or a fine up to a maximum of £5000, or both. The court also has the power to grant a compensation order, payable by the guilty party to the consumer, either in addition to the fine or instead of it.

The Need for Records

Regulation 8 concerns 'producers' operating systems. Customers now have the right to be supplied with all relevant product information and specifications to enable the consumers to be aware of possible risks with the product. If dangers relating to the product are discovered by the producer then the producer must be able to inform the customers, as soon as is practicable, of the dangers in order that this requirement can be carried out expeditiously it is essential that customer records are accurate and complete. Quality practitioners would be well advised to familiarize themselves with the requirement of Regulation 8 which state that:

1 Within the limits of his activity, a producer shall :
   (a) provide consumers with the relevant information to enable them to assess the risks inherent in a product throughout the normal or reasonably foreseeable period of its use, where such risks are not immediately obvious without adequate warnings, and to take precautions against those risks; and

(b) adopt measures commensurate with the characteristics of the products which he supplies, to enable him to be informed of the risks which these products might present and to take action, including, if necessary, withdrawing the product in question from the market to avoid those risks.

2 The measures referred to in sub-paragraph (b) above may include, whenever appropriate:
   (i)   marking of the products or product batches in such a way that they can be identified;
   (ii) sample testing of marketed products;
   (iii) investigating complaints; and
   (iv) keeping distributors informed of such monitoring.

Requirements of Distributors

Regulation 9 is aimed directly at distributors by imposing upon them a *duty of care* to ensure that no producer places on the market a product which is unsafe. Indeed it would seem that this particular regulation places an onerous responsibility upon the distributor, in that it refers to knowledge or presumption when it states in 9(a): 'a distributor shall not supply products to any person which he knows, or should have presumed, on the basis of the information in his possession and as a professional, are dangerous products.'

Great care needs to be taken in this area. It would be quite wrong to assume an 'ignorance is bliss' attitude will provide a successful defence. The references to the presumption of knowledge and professionalism suggests quite clearly that the distributor should ensure that they are fully aware of the product's safety together with its conditions of use etc., before placing it on the market. Indeed this a specific area where supplier quality assurance and the control of incoming product, together with the necessary supporting documentation, stand out firstly as necessary preventative measures and secondly as possible aids to a legal defence.

Further responsibility is placed upon distributors by Regulation 9(b) which states: 'within the limits of activities, a distributor shall participate in monitoring the safety of products placed on the market, in particular by passing on information on the risks and co-operating in the action taken to avoid those risks.'

The limit to which a distributor is required to participate in the monitoring of safety will depend upon the activities and products involved and will as a result vary between different distributors and different types of products. What should be clear is the need for distributors to be in a position to make themselves, and as a result their customers aware of any information which exists regarding the risks involved with the products being supplied.

The Use of Standards

The use of standards and codes of practice and the degree of reliance which can be placed upon them is specified in Regulation 10 (1):

'Where in relation to any product such product conforms to the specific rules of the law of the United Kingdom laying down health and safety requirements which the product must satisfy in order to be marketed there shall be the presumption that, until proved to the contrary, the product is a safe product.'

Note that as with the Consumer Protection Act 1987 the requirement placed upon the prosecution – failing to comply with these Regulations being a criminal matter – is to prove that the product is unsafe. Regulation 10 (2) further states:

'Where no specific rules as are mentioned or referred to in paragraph (1) exist. the conformity of a product to the general safety requirement shall be assessed taking into account:

(i) voluntary national standards of the United Kingdom giving effect to a European standard; or
(ii) Community technical specifications; or
(iii) if there are no such voluntary national standards of the United Kingdom or Community technical specifications:
   (aa) standards drawn up in the United Kingdom; or
   (bb) the codes of good practice in respect to the health and safety in the product sector concerned; or
   (cc) the state of the art and technology;
   and the safety which consumers may reasonably expect.'

Note that this Regulation does not mean that a product is absolutely safe if it adheres to a particular standard, what it does say is that conformance with such standards shall be taken into account.

Defence of Due Diligence

Similar to a number of pieces of legislation, the General Product Safety Regulations provide for the defence of due diligence. Regulation 14 states that:

'. . . it shall be a defence for that person [the defendant] to show that he took all reasonable steps and exercised all due diligence to avoid committing the offence.'

The phrase to '. . . show that he took all reasonable steps . . .' seems to automatically suggest that the defendant would be able to provide evidence that some form of meaningful procedures existed at the time of the alleged offence, and that any such procedures had been adhered to. That being the case and irrespective of the product in question, there seems to be a number of areas for which there is a need for a well constructed and satisfactorily operated quality assurance management system.

Regulation 14 goes on to provide the defendant with the opportunity to establish that any offence which has been committed was due to:

(a) '. . . the act or default of another . . .', or
(b) '. . . reliance on information given by another . . .'

In either of these two situations it is a requirement of the Regulations that the defendant must inform the prosecution, no later than seven days prior to any hearing of the proceedings. In taking this course of action Regulation 14(3) requires that the defendant '. . . shall give such information identifying or assisting in the identification of the person who committed the act or default . . .'.

Here again there would seem to be an obvious need for the defendant to be able to provide documentary evidence of the source of the defence. However, Regulation 14(4) makes it an additional requirement for the defendant to show that: 'it was reasonable in all the circumstances for him to have relied on all the information, having regard in particular:

(a) to the steps which he took, and those which he may reasonably have taken, for the purposes of verifying the information; and
(b) to whether he had any reason to disbelieve the information.

Time Limit

One final consideration concerns the time period involved. The Regulations (16) stipulate that a case must be brought before a Magistrate within twelve months of the date of the offence.

Controlling the Risks

The combination of the risks created by contractual obligation, the duty of care owed with regard to negligence, and the Consumer Protection Act, present producers with a formidable package of product liability law with which they have to contend. In order to control and minimize the multi-faceted risks requires a multi-functional approach should be an integral part of the business's quality assurance procedures. The activities referred to can be categorized as:

1 Risk analysis or assessment.
2 Risk transfer.
3 Risk reduction.

Risk Assessment

This is in essence a product liability audit, undertaken in order to establish the identity and the extent of any risks which may occur during the normal use of the product. Risks need to be categorized and evaluated in terms of their seriousness, in order that plans may be made to either eradicate or minimize their effect. The methods used to perform risk assessment will be largely dependent upon the type of product or the industry concerned. Ideally expertise should be gathered together from the areas of marketing, design, production, inspection, service and of course quality assurance, to establish how the product could be harmful if it were subsequently used in certain ways and under certain conditions. Needless to say, potential, problems which might occur due to the production process as a whole need to be established and where possible eradicated. The use of 'brain storming' is a technique which can be used irrespective of the type of industry, product or service, to establish the nature of the risks involved. Failure mode and effects analysis (FMEA) could then be used to establish the probability of occurrence and the severity of such risks, in order to create a priority for attention based upon the seriousness of the risks in question. The immediate problem that such an approach appears to offer is the costs involved. FMEA is a painstaking process which, if it is to be performed properly, requires the use of a good deal of expertise with regard to the process or service being analysed. A somewhat short-sighted view given the costs involved in fighting a legal action, let alone the penalties of being found to be at fault. The alternatives which companies may decide upon could be:

1 not to pursue investment in the development of new product areas; or
2 to limit the market to which the companies' products are offered.

The economics of either may well prove to be prudent with regard to the potential cost of possible consumer protection litigation. However, such a head in the sand approach could equally prove to be economically unviable in a consumer economy that is consistently looking for innovative and constantly improving products and services.

Risk Transfer

The production of consumer goods usually necessitates a chain of supply, with a similar situation arising in the service sector. For example, with regard to manufactured products, the chain may well be as shown in Figure 4.2.

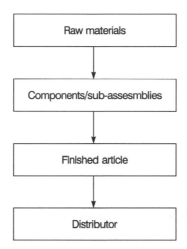

Figure 4.2 Supply chain – manufactured goods

Similarly, a supply chain for the tourist industry might look like that shown in Figure 4.3 which whilst appearing to be quite different from manufactured articles still equates to the groups as raw material, component/sub-assembly suppliers, final producers and distributors.

In such circumstances it may be possible for one organization to transfer the risks to another using carefully constructed contracts; e.g.. a manufacturer may require that raw materials, components, or sub-assemblies are:

(a) produced to relevant manufacturing, and/or quality standards;
(b) supplied with a certificate of conformance;
(c) covered by a warranty which will indemnify the manufacturer against claims.

117

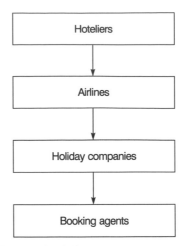

Figure 4.3 Supply chain – service industry

Whilst the inclusion of (c) is good business practice, both (a) and (b) should be included as a matter of form within the purchasing procedures of any good quality assurance system. It is worth reiterating at this point that, what is not permissible under the Consumer Protection Act 1987, is for companies to use contract terms to disown liability. To do so may also in some circumstances also be contrary to the Unfair Contract Terms Act.

In addition to these three points, it needs to be borne in mind that the Consumer Protection Act imposes both joint and several liability. This means that one defendant company may be required to meet all the claims for compensation in cases where the co-defendants are unable to do so. Ensuring that suppliers are adequately insured therefore, should also be part of the quality procedures involving the evaluation of sub-contract suppliers.

For risk transfer to be fully effective it is absolutely fundamental that good documentation and its control is incorporated into the quality system, in order to establish the points at which liabilities start and finish.

Risk Reduction

Risk reduction activities should be an integral part of the quality assurance management system, involving the majority, if not all areas of the company's activities. A programme of activities should be created to ensure that each department plays an active role in risk reduction as a part of their normal every day functions. Procedures and techniques which could be included on a department basis are:

1 Design
  (a) Design reviews should permit the inclusion of expertise to ensure that the safety of the product is correctly assessed prior to the design's acceptance.
  (b) Failure mode and effects analysis be instigated to assess the probability of risk occurring.
  (c) Standards and codes of practice be used, and where necessary expertise sought from the suppliers of bought out components.
  (d) Adequate research and development be undertaken to ensure the safety of the finished product.

2 Production
  (a) Implement a policy of defect prevention rather that defect detection.
  (b) Instigate and maintain procedures for traceability and recall.
  (c) Ensure that the procedures which have been implemented as part of the quality system are adhered to, and just as importantly, modified when areas of concern are highlighted.

3 Purchase
  (a) Assess and verify all suppliers to ensure they are capable of meeting the necessary quality, and product safety requirements.
  (b) Have suppliers instigate failure mode and effects analysis as part of their contractual obligations.
  (c) Ensure that all contracts specify the liabilities incurred by the parties involved. and that suppliers have the necessary degree of insurance to ensure that they are capable of meeting all the costs and damages which may be incurred due to litigation.

4 Service
  (a) Analyse all customer feedback, and highlight safety related situations as they are reported.
  (b) Ensure corrective action is taken expeditiously when the need arises.
  (c) Service manuals should be kept up to date, with service personnel being specifically advised on all changes which may reflect on matters of safety.
  (d) Ensure the quality of replacement parts.

5 Training
  (a) Ensure that personnel involved with design are fully appraised of relevant technological development. Given the pace of advancement in some industries, the 'state of the art defence' could well disappear rapidly for any given situation.
  (b) Ensure that appropriate training is undertaken by personnel involved with all product safety related issues.

    (c) Ensure that all involved personnel are trained in the preparation of reports and documentation, especially that which may be relevant in the defence of product liability litigation.

6 Sales and marketing

    (a) Where possible ensure the limitations of a product are clearly highlighted.

    (b) Take extreme care when making claims regarding the use to which a product may be put. It needs to be understood that such claims, although made verbally, could well be:

       (i)   proven in a court of law to constitute part of a contract;

       (ii) in breach or the Trade Descriptions Act.

The above precautions should be part of the procedures of the quality system. However, it needs to be borne in mind that the unexpected is exactly that, and as a matter of course product liability insurance should be taken out by all producers.

Case Study

The following case study contains three separate scenarios each of which is centred around the core information given at the beginning. The task for the reader is to try to determine the responsibilities, penalties and probable outcomes of litigation, for each scenario, based upon the Consumer Protection Act 1987. In addition the reader is asked to try to suggest what preventative action could have been taken. The authors' explanations are given at the end of the chapter.

Core information

Mrs A. purchased an electric toaster from an electrical retailer as part of a closing down sale. On the toaster was a plastic moulding which acted as the retaining clip for the electrical cable. During the course of normal use, the moulding broke away, and as a result the live wire became detached from its terminal. Unaware that this had happened the toaster was kept in domestic use. Subsequently Mrs A's husband received an electric shock which caused him to fall and break his arm. The incident occurred seventeen months after the toaster was purchased, and the guarantee for the product had expired five months previously. Mr A consulted a solicitor very shortly after the incident.

Scenario One

Upon examination of the toaster the plastic clip was discovered to be faulty because of, what was described as an inherent design fault. The clip was part of a design project awarded to a sub-contract design office who were not part of the manufacturer's organization.

Scenario Two

The broken moulding was attributed to its retaining screw being over tightened during assembly at the manufacturer's premises. The actual reason for this happening was never discovered, as the toaster in question had been produced over a six year period which had finished six months prior to the date of the faulty toaster's sale. The company's production engineers did suggest a number of theories, although they were unable to agree as to the actual cause. What they did agree to was that due to the construction of the toaster the retaining screw in question was inaccessible without dismantling the toaster's casing and there was no evidence to suggest that this had been done.

Scenario Three

The fault was caused by the plastic moulding which was one of a batch which had been produced from an incorrect compound. The moulding was produced by a sub-contract plastic moulding company and supplied by them together with a certificate of conformance in accordance with the toaster manu-facturer's quality procedures.

Authors' explanation – core information

Mr A would be making a claim based upon the evidence that the electric toaster proved to be unsafe in the course of normal use, and that as a result he had suffered 'damage'; specifically an electric shock and the ensuing broken arm. As each of the following scenarios accepts that there was a fault in the product, Mr A's first decision would be to decide exactly whom he should make his complaint against. An important point here is that the complaint is Mr A's. Although his wife made the actual purchase it is Mr A who is seeking redress for damage suffered due to a faulty product. Prior to the Consumer Protection Act 1987, as he was not the purchaser and therefore had no contractual rights with regard to the toaster, his only course of action would have been to try to establish that there had been an act of negligence which had led to his injury.

The retailer having gone out of business does not absolve any other party from liability, which as far as the Consumer Protection Act is concerned is both joint and several. The manufacturer in this case would seem to be the best place for Mr A to direct his law suit. The concerns over the expiry of the guarantee period are irrelevant in this case. Manufacturers guarantees are offered as an addition to a consumer's statutory rights not in place of them.

Finally, Mr A has a period of up to three years in which to start his action. Given that the case study states that he consulted a solicitor shortly after the event, it can be safely assumed that Mr A's action would not be barred on that point by any Act of Limitations. As to whether or not he would be barred from pursuing his claim due to any other time limit will become apparent from each of the scenarios in turn.

Explanation – Scenario One

The reference to an inherent design fault, regardless of where the design was produced, indicates a failure to conduct design review procedures on the component in question. Given the nature of the component was it identified as being a 'safety critical' part ? If so what precautions were taken? Was a failure mode and effects analysis carried out? If not why not? If so at what point did the analysis fail to identify what became the subsequent problem? These are the types of questions which need to be answered if this type of problem is to be prevented from occurring in the future. Sadly, experience tends to suggest that the first concern will be the attempt to off set litigation rather than to review any shortfall in quality procedures. The point that would therefore be overlooked is that if such procedures existed and the facilities and expertise existed for their use this may help the company in trying to establish that, as far as the company was concerned, all due diligence had been exercised. The real point to be made is that if such procedures did exists and were exercised correctly the problem would have been avoided in the first place.

From the point of view of deciding whom to seek damages from, the authors suggest that Mr A would be best advised to direct his efforts at the manufacturer. The simple economics are such that a sub-contract design office, with some notable exceptions, tends to be a small business, with limited assets and as a result quite unable to meet the financial demands of a civil settlement. Given that this leaves the manufacturer with the financial responsibility it would once again seem to highlight that they themselves had good supplier quality assurance procedures which involved sub-contract design as well as component and raw material suppliers.

Explanation – Scenario Two

In this situation it has been established that the assembly was responsible, although the circumstances which caused the fault were never clearly established. The company's only defence would be to establish that the toaster did not have a defect at the relevant time, in other words it had been had been tampered with after it had left their premises, or that the toaster had been misused – not used for the purposes for which it had been intended. Reading the scenario shows that it had been established that there was no evidence to substantiate such a claim. Examining the time scales involved shows that the claim being made by Mr A is well within the ten year time frame permissible under the Act. This would be the case even if the toaster in question had been the very first one of its model to be made.

In so much as the Consumer Protection Act 1987 is concerned, in this scenario the manufacturers would appear to be liable, which would leave the courts to decide upon the scale of damages to be awarded and whether or not Mr A's legal costs should also be met by the manufacturer.

For the manufacturer the penalties do not end there. Additional costs could include:

1 The costs of trying to establish as to whether or not this was an isolated incident.
2 If it wasn't, the cost of recalling additional toasters.
3 The possibility of further legal action from other purchasers who may be 'damaged' by faulty toasters still in domestic use.
4 The possibility of prosecution either under the Consumer Protection Act or under other applicable legislation.

In addition and the most difficult to quantify is the costs involved due to the damage sustained to the manufacturer's reputation, which may reflect on the scale of future sales.

Possible preventative measures

The first area of concern is the inability of the production engineers to establish the actual reason for the retaining screw being over tightened. This alone suggest that there was poorly maintained documentary evidence with regard to procedures which obviously affect quality. That not withstanding, the most logical approach would be to start by reviewing the design, then examining the process control, and checking the inspection records and procedures for the product involved. Finally would come an investigation of the assembly workers involved and an examination of their training records.

Taken in the sequence suggested: firstly, if poor design control been the answer then there would surely have been a drawing which stated the maximum torque to which the screw should have been tightened. Had that been the case, providing the designs were still in existence, and they most certainly should have been, then there would of been documentary evidence to establish the cause. Was poor design review to blame? Was it carried out correctly, or indeed, is there a clear and logical design review procedure?

Secondly, the process controls involved give rise to a number of questions:

1 Did documented procedures exist at the time the faulty toaster was produced?
2 Did the documentation clearly specify the methods of assembly to be used?
3 Was suitable equipment in use, was it correctly set, and in keeping with that, did it perform to its current test status?

Thirdly, inspection. The same questions arise with regard to the existence of procedure, equipment and methods used. The question that is yet to be asked is whether or not the part of the assembly which gave rise to the problem was actually inspected? Was this something that was the responsibility of the assembly worker concerned? There is no reason why it should not be providing: that in the first instance the person involved can be identified. Are the products involved tagged, or stamped, or otherwise identified against the assembly worker involved? The reasons for these questions are not so that blame can be apportioned, but rather so the company can investigate the fourth area of concern. Was the assembly worker sufficiently well trained and competent to carry out his/her function? If so is there a record of this, and if not what procedures exist to remedy this lack of training?

These questions are not suggested as the means for conducting a witch hunt, but rather as a list of preventative measures that should have been taken to prevent the incident in the first place and failing that, to ensure that it is never repeated. Good quality assurance procedures should reflect efficient, effective and safe methods of working which ultimately are reflected in the production of safe products. Time spent on those above could well have prevented the circumstances which arose in Scenario 2. They could definitely help to ensure that the same mistakes are not repeated.

Explanation – Scenario 3

As liability under the Consumer Protection Act is both joint and several, Mr A is presented with the option of suing either the manufacturer, the sub-contractor, or both. The suppliers of raw materials or component parts having an equal 'duty of care' to the end user. For the purposes of this scenario it can be assumed that Mr A has been advised by his solicitor to enter into litigation against the manufacturer. This may seem to leave the manufacturer in an unfair situation as they did not actually manufacture the defective cable clip. Unfortunately for the manufacturer, that fact does not leave them with a defence in law. Since the defective part was installed into the toaster at their premises the defect is said to exist at the 'relevant time' – the time it was supplied by the manufacturer. As a result the manufacturer stands liable to all the penalties and costs mentioned in the explanation to Scenario 2.

However, the manufacturer does have a source of redress. As the sub-contractor provided the cable clip, which subsequently was revealed to have been produced from an incorrect compound, the sub-contractor had in fact failed in their contractual obligations to the manufacturer. Given the penalties which the manufacturer would undoubtedly incur, it would seem highly likely that the manufacturer would seek compensation from the sub-contractor.

It would appear that from a quality assurance point of view that the manufacturer did have certain quality assurance procedures in operation. The sub-contractor had provided the components together with a certificate of conformance. Such certification is normally only supplied at the request of the customer (in this case the manufacturer), and its existence suggests an understanding of the principles of risk transfer. However, the incident highlights areas of the quality assurance system where the procedures may well be improved. Possible improvements in such areas as:

1 Evaluating the performance and capabilities of sub-contractors in general and the plastic supplier in particular.
2 Examining the verification activities that take place at sub-contract companies.
3 The re-evaluation of levels of inspection that the sub-contractors are required to use and, specifically with regard to this scenario, the inspection procedures which are use to establish that the correct raw materials have been used.
4 The possibility of revising, or if necessary installing goods inwards inspection procedures.

If the manufacturers in this scenario are successful in seeking compensation, they will still not escape from the incident without cost. The trouble involved, and the loss of reputation could well be substantial. It would seem reasonable if the lessons were learned and the quality procedures improved where necessary. It would certainly be in keeping with the generally accepted quality assurance philosophy of continuous improvement.

As an addendum to these scenarios, the parties involved should also be aware of their responsibilities under the General Product Safety Regulations 1994. During the incident Mr A actually suffered injury, a fact which could leave the party responsible facing a criminal prosecution.

5

Sale and Supply of Goods Legislation

In Chapter 2 on contract law, it was first explained that a contract is an agreement between two or more persons which is enforceable by law. One of the first comments which must be made regarding the legislation on the sale of goods, is that it will only apply where there is a contract in force, and only the parties to the contract can sue for breach of that contract. Looking back again to the chapter on contract law, it is explained that the Sale of Goods Act 1979 (as amended by the Sale and Supply of Goods Act 1994), implies terms into contracts for the sale of goods which entitle a party to the contract to sue for breach of contract if any of these implied terms are breached.

Contract law was also said to impose strict liability on the parties to the contract. This means that a party to the contract can be liable for breach of contract regardless of whether they are to blame for the breach or not. Negligence is fault based liability and contract is liability without fault (strict liability). This can be demonstrated by re-examining the case of:

Grant v. *Australian Knitting Mills* (1936)

Mr Grant was the man who bought some long-johns from a store. He developed very severe dermatitis because the manufacturers failed to remove sulphur from the underwear during the manufacturing process. The manufacturers were found to have been negligent (at fault) in allowing the sulphur to remain in the garments. The store (the sellers) were also found liable. This time however, the liability arose through the contract of sale Mr Grant had with the store. Although the store was in no way to blame for supplying faulty garments, they were still liable under the Sale of Goods Act 1893 for selling goods which were not of 'merchantable quality' (s.14(2)) and for selling goods which were not 'fit for their purpose' (s.14(3)).

It may be considered at this stage, especially by those readers involved in the retail business, that there is little point in explaining about legislation that imposes liability for something one has no control over. No quality assurance management system can prevent liability from arising, because the retailer has to do nothing wrong in order for liability to arise. To some extent this is a valid argument. However, if the legislation and common law is examined, benefits from establishing certain quality assurance measures into the workplace will soon become apparent.

The Sale of Goods Act 1979 (as amended by the Sale and Supply of Goods Act 1994) is not the only legislation that implies terms into contracts. The Supply of Goods and Services Act 1982 (as amended) implies terms into contracts where goods are 'transferred' under a contract, as opposed to being 'sold' under a contract and where there is a contract for a service to be performed. Hire-purchase agreements are covered by the Supply of Goods (Implied Terms) Act 1973 as amended by the Consumer Credit Act 1974. There are also a number of other statutes which affect day to day commercial and consumer transactions which are outside the scope of this book. This chapter is going to take a brief look at the legislation relating to the sale of goods and some service contracts. It can now be seen that it is not only retailers who are affected by the legislation, but also persons who carry out service contracts. As will be shown later, retailers include those retailers selling second hand goods as well as those selling brand new goods. Private sellers may also incur liability for breach of some implied terms.

Sale of Goods Act 1979

The following sections will discuss in turn definitions under the Act.

Sale

Section 2(1) of the Act defines a contract of *sale* of goods as:

> a contract by which the seller transfers or agrees to transfer the property in the goods to the buyer for a money consideration, called the price.

Subsections 4 and 5 further define these transactions.

> Section 2(4): Where under a contract of sale the property in the goods is transferred from the seller to the buyer the contract is called a sale.

> Section 2(5): Where under a contract of sale the transfer of the property in the goods is to take place at a future time or subject to some condition later to be fulfilled the contract is called an agreement to sell.

So, in order for the Sale of Goods Act 1979 to apply, there must be a contract of sale. Distinctions must be made between contracts of hire-purchase, which are covered by separate legislation, and contracts for the supply of goods and services, which are also covered by separate legislation.

Goods

Section 61 of the Act defines goods as including 'all personal chattels other than things in action and money'. 'Emblements, industrial growing crops, and things attached to or forming part of the land which are agreed to be severed before sale or under the contract of sale', are all included in the above terms. The definition is very broad and covers a wide range of 'things'. Things in action are excluded from the Act, which means that company shares (as non-physical things) are excluded.

Implied terms

All the formalities of contract law are required before a valid contract exists. These are, offer, acceptance, intention to create legal relations, consideration and capacity, as discussed previously. Once the formalities have been completed and a sale of goods has taken place, then the implied terms of the Sale of Goods Act 1979 become part of that contract and the parties are bound by not only the express terms, agreed to by the parties, but also the implied terms of the statute.

Title to goods

Section 12 implies a condition on the part of the seller that in the case of a sale he has a right to sell the goods, and in the case of an agreement to sell he will have such a right at the time when the property is to pass.

Section 12 applies to all contracts of sale. This includes sales by private sellers as well as those who are selling in the course of a business.

Rowland v. *Divall* (1923)

Divall sold Rowland a car for the sum of £334. Divall had bought the car from someone who had stolen it. Rowland had to return the car to its rightful owner, after having had the car for four months. Rowland then sued Divall under s.12 of the Sale of Goods Act 1979 for breach of the implied condition that the seller (Divall) had the right to sell the car. As Divall did not have good title to the car, he could not pass good title on to the purchaser. Rowland had paid £334 for the ownership of the car, but as he did not acquire rights of ownership there was a total failure of consideration on Divall's behalf. Rowland was therefore entitled to the full purchase price of £334.

The effect of strict liability on a seller under section 12 of the Act should encourage persons buying goods to check that the person selling those goods has good title (legal ownership) to the goods. A retailer, for example, needs to be sure that his supplier is reliable and that the retailer has proof of the seller's legal ownership before buying any goods from that supplier. The effect on a business of having bought goods from a person without good title to those goods can be detrimental in more ways than one. Firstly, if the business still has possession of the goods and they then have to be returned to their rightful owner, the business will be out of pocket as there will be no goods and the purchase price may already have been paid to the seller of those goods. Secondly, if the goods have been sold on to an innocent third party when they have to be returned to the original owner, then the likelihood is that the business will be sued for breach of s.12 of the Sale of Goods Act 1979 by that third party. Dealing with reputable suppliers and the keeping of full and accurate records of goods purchased can have definite advantages for the business.

1 If suppliers are reputable then situations such as this are less likely to happen.
2 If accurate records are kept, then the retailer himself can sue his supplier for breach of the implied condition as to title. The retailer can sue either to recover the price paid, where he himself has had to give up the goods, or to be indemnified for having paid out the purchase price to the third party, who has sued the retailer for breach of contract.

Sale by description

Section 13 (1): Where there is a contract for the sale of goods by description, there is an implied condition that the goods will correspond with the description.

S13 (2): If the sale is by sample as well as by description it is not sufficient that the bulk of the goods corresponds with the sample if the goods do not also correspond with the description.

S13 (3): A sale of goods is not prevented from being a sale by description by reason only that, being exposed for sale or hire, they are selected by the buyer.

This section applies to both private and commercial sales, as demonstrated by the following cases:

Arcos Ltd v. *E A Ronaasen & Son* (1933)

A quantity of wooden staves, described as 'half an inch thick', was sold by the seller to the buyer for the purpose of making barrels. All but 10% of the staves were between half an inch and five-eighths of an inch thick. The remaining 10% were found to be over five-eighths of an inch. All the staves were suitable for making barrels, but the buyer was held to be entitled to reject all the staves as they did not match the description given.

Beale v. *Taylor* (1967)

The plaintiff saw an advertisement for a 1961 Triumph Herald. He viewed the car and purchased it from the defendant. He took the car to a garage some weeks later, after experiencing some problems with it. He discovered that the car was, in fact, two different models which had been welded together. The rear half was a 1961 Triumph Herald, but the front half was an earlier model. The plaintiff sued the defendant for breach of the implied condition that the goods must match their description. Even though he had inspected the car before buying it, he was still held to be entitled to damages for breach of contract as he had relied partly on the description given.

A breach of the implied term in section 13 gave the buyer the right to reject the goods. This was because there had been a breach of a condition of the contract. In Chapter 2 on contract law it was explained that a breach of a condition of a contract is more serious than a breach of warranty. A breach of a condition is a breach of a fundamental term which goes right to the root of the contract. The breach of a condition entitles the innocent party to treat the contract as terminated and to claim damages in compensation. However, it was considered that not all breaches of the implied terms under the 1979 Act were so serious that the buyer should be entitled to reject the goods, as this appeared very harsh where the breach was only slight. Because of these considerations the Sale and Supply of Goods Act 1994 has amended the Sale of Goods Act 1979. The effect of one of these amendments can be seen in section 15A which the 1994 Act introduced into the 1979 Act.

Section 15A provides that where a seller can show that the breach of ss.13–15 is so slight that it would be unreasonable for a non-consumer buyer to reject, the breach is to be treated as a breach of warranty and not a breach of condition.

The effect of s.15A differs as to whether the buyer is a consumer or not. Where the buyer is a consumer, the right to reject the goods for any breach of the implied terms of the 1979 Act has been retained. A commercial buyer, however, will be prevented from rejecting goods where the breach is so slight that it would be unreasonable to reject.

Quality of the goods

Section 14 of the 1979 Act provided that where a seller sells goods in the course of a business there is an implied condition that the goods supplied are of 'merchantable quality'. The 1994 Act has amended this to 'satisfactory quality'. Section 14(2)A provides that goods are of satisfactory quality if they meet the standard that a reasonable person would regard as satisfactory, taking into account any description of the goods, the price (if relevant) and all other relevant circumstances.

Section 14(2)B provides that the quality of goods includes their state and condition and the following are aspects of quality:

(a) fitness for all purposes for which goods of the kind in question are commonly supplied;
(b) appearance and finish;
(c) freedom from minor defects;
(d) safety; and
(e) durability.

Section 14 differs from sections 12 and 13 because it only applies where the seller sells goods in the course of a business. If goods are bought from a private seller then the provisions of s.14 do not apply.

The case law decided under the old definition of 'merchantable quality' is still useful in determining how the courts will decide issues regarding satisfactory quality, until cases have been decided on the new requirement of 'satisfactory quality'.

As will be seen by the following cases, section 14 applies to the sale of brand new and second hand goods.

Shine v. *General Corporation Ltd* (1988)

A motoring enthusiast bought a second-hand specialist sports car. He later discovered that the car had been involved in an accident and submerged in water for 24 hours. He then rejected the car under s.14(2) claiming the car was not of merchantable quality. The Court of Appeal held that the car was not of merchantable quality. Bush J said that no member of the public who was aware of the car's history, '*would touch it with a barge pole unless they could get it at a substantially reduced price to reflect the risk they were taking*'.

Bartlett v. *Sidney Marcus Ltd* (1965)

A car dealer sold a second-hand car to the plaintiff. The dealer told the plaintiff that the car had a defective clutch and a new price was negotiated by the parties to take account of this defect. The repairs to the clutch turned out to be far more expensive than the plaintiff had anticipated when he bought the car. He claimed the car was not of 'merchantable quality' under s.14(2) and sought to reject the car. The court held that the car was of merchantable quality and it should be expected that a second-hand car would have defects sooner or later.

Rogers v. *Parish Ltd* (1987)

The plaintiff bought a new Range Rover for £16,000. The vehicle proved to be unsatisfactory and after only a few weeks the dealers exchanged the Range Rover for a new one. The new vehicle proved to be just as much trouble as the first. The bodywork was defective, there was excessive noise coming from the gearbox and the engine misfired. The plaintiffs sought to reject the vehicle under s.14(2) because it was not of merchantable quality. The court held that the vehicle was not of merchantable quality. Although the car was capable of getting from A to B, a reasonable person spending £16,000 on a vehicle would expect it to look good, handle well, and provide a high degree of comfort and reliability. The Range Rover failed to live up to these expectations.

These cases highlight the factors that the courts took into consideration when determining whether goods were of merchantable quality. The aspects of quality listed in s.14(2B) by the amendments of the Sale and Supply of Goods Act 1994, regarding satisfactory quality, reflect these factors.

There are two situations in s.14(2C) which can result in the buyer losing his right to complain. The first is where the seller points out the defect to the buyer (as in the case of *Bartlett* v. *Sidney Marcus*) and the second is where the buyer decides to check the goods and fails to notice the defect.

Fitness for a particular purpose

Section 14(3) of the 1979 Act provides that where a seller sells goods in the course of a business and the buyer makes known to the seller any particular purpose for which the goods are being bought, there is an implied condition that the goods supplied are reasonably fit for that purpose, whether or not that is a purpose for which such goods are commonly supplied, except where the circumstances show that the buyer does not rely, or that it is unreasonable for him to rely, on the skill or judgment of the seller.

Again, s.14(3) only applies to goods sold in the course of a business and not to private sales. A case which has already been considered before gives an illustration of goods being fit for their purpose.

Grant v. *Australian Knitting Mills* (1936)

Mr Grant sued the manufacturers in a negligence action and sued the retailers for breach of the implied terms under s.14(2), as the long-johns he purchased were not of merchantable quality, and also under s.14(3) as they were not fit for the purpose. Mr Grant did not have to expressly state the purpose for which the long-johns were being bought as it was clear by implication that he bought them to wear. He succeeded in all three actions.

Contrast the decision in the following case.

Griffiths v. *Peter Conway Ltd* (1939)

The plaintiff, who had an abnormally sensitive skin, bought a Harris Tweed coat from the defendants. Within a short period of time, she had contracted dermatitis from the coat. The defendant retailers were not liable under s.14(2) as the coat was of merchantable quality, because persons with normal skin would not have been affected. The retailers were not liable under s.14(3) either, as it was fit for the purpose and the plaintiff had not told the defendants of her abnormally sensitive skin when she bought the coat. Even if the plaintiff had advised the shop that she had particularly sensitive skin, and the shop had sold her the same coat, it is arguable that the shop would still not have been liable. Section 14(3) will not apply where:

(a) the buyer does not rely on the skill and judgement of the seller; or
(b) it was unreasonable for him to rely on it.

The retailers may claim that it was unreasonable for the buyer to rely on their skill and judgement when buying the coat, in these particular circumstances, as they do not have any knowledge of skin conditions.

Sale by sample

Section 15 of the 1979 Act provides that:

Section 15(1): A contract of sale is a contract for sale by sample where there is an express or implied term to that effect in the contract.

Section 15(2): In the case of a contract for sale by sample there is an implied condition:

(a) that the bulk will correspond with the sample in quality;
(b) that the buyer will have a reasonable opportunity of comparing the bulk with the sample;
(c) that the goods will be free from any defect, rendering them unmerchantable, which would not be apparent on reasonable examination of the sample.

This section, like ss.12 and 13, applies to both business and private sales. A case which illustrates the application of s.15 is given below.

Godley v. *Perry* (1960)

A six year old boy bought a plastic catapult from the defendant shopkeeper, Mr Perry. The catapult broke when used and the plaintiff lost an eye. The plaintiff sued the shopkeeper for breach of the implied conditions in ss.14(2) and 14(3). The plaintiff succeeded in recovering damages from the defendant as the catapult was not of merchantable quality, or fit for the purpose for which it had been supplied. The defendant had bought the catapult from a wholesaler in a sale by sample. He had tested a catapult from the sample, by pulling back the elastic, but no fault had been revealed. The defendant sued the wholesaler for breach of the implied condition under s.15 and succeeded. In turn, the first wholesaler sued the second wholesaler, who had sold the catapults to him by sample, for breach of the implied terms under s.15.

This is still a case of strict liability, as the person sued was not to blame for the faulty goods. However, each person in the contractual chain can sue the person they purchased the goods from, either for breach of s.15 where there was a sale by sample, or for a breach of ss.14(2) and/or (3) where the goods are not of satisfactory quality or fit for the purpose for which they were supplied. Eventually, the last person to be sued will be the manufacturer who will be the person at fault for producing faulty goods.

Manufacturers need a quality assurance system in place to reduce the chances of mistakes being made in the manufacturing process, and to ensure they are detected and rectified before leaving the manufacturing plant. Businesses that are buying goods for the purpose of selling the goods on, need to ensure that their quality assurance systems include keeping accurate record of names and addresses of suppliers and details of goods purchased from each. This ensures that the relevant information is readily available if it becomes necessary to take action against suppliers for supplying goods which:

1 do not match their description;
2 are not of satisfactory quality;
3 are not fit for the purpose for which they were supplied; or
4 did not match the sample, in a sale by sample.

Dealing with reputable suppliers can be vital to the success of any action taken. It is only worth suing another party if that party is solvent and has sufficient funds, or insurance, to pay out the damages claimed. The danger of dealing with an unknown party is that it is not known whether they deal with reputable suppliers themselves, whether the goods are of satisfactory quality, or how profitable the business is. The one problem with relying on the ability to be indemnified by a supplier, for any action taken by the buyer of goods, is that it only works successfully if the contracting party is solvent and the address of the supplier is up to date and accurate. The up-dating of records should also form part of a quality system. Also, if one supplier in the chain of suppliers is no longer in business, then the contract chain will be broken, the claims will stop and the manufacturer will not be the party ultimately to pay damages for the defect.

The Supply of Goods and Services Act 1982 (as amended by the Sale and Supply of Goods Act 1994)

The Supply of Goods and Services Act 1982 is divided into two parts. The first part deals with the implied terms in contracts for the supply of goods and the second part deals with the implied terms in contracts for services. Contracts for the supply of goods cover two situations in the Act. Firstly, sections 2–5 of the 1982 Act imply terms into contracts for work and materials, where the ownership of goods passes from one party to another. An example of this type of contract could be where there is a contract for building work to be carried out. The building materials will become the property of the commissioner of the work and the building work itself is also subject to the provisions of the Act. Sections 2–5 have broadly the same effect on these contracts as sections 12–15 of the Sale of Goods Act 1979.

Section 2 1982 Act is concerned with implied terms as to title of the goods. In a contract for the transfer of goods, there is an implied condition that the transferor has the right to transfer the property.

Section 3 provides that in a contract for the transfer of goods, the goods transferred must match their description.

Section 4 concerns the quality of the goods transferred. In a contract for the transfer of goods there is an implied condition that the goods are of satisfactory quality. There is no such condition where the defects have been specifically brought to the attention of the transferee at the time of the contract, or where the transferee examines the goods before the contract is made and the defects should have been revealed by that examination. Also, under a contract for the transfer of property where the transferee expressly or by implication makes known to the transferor any particular purpose for which the goods are being acquired, there is an implied condition that the goods transferred are reasonably fit for that purpose whether or not that is a purpose for which such goods are commonly supplied. This does not apply where the transferee does not rely on or it is unreasonable for him to rely on the skill or judgement of the transferor.

Section 5 concerns transfers by sample. This section implies conditions that the bulk will correspond with the sample in quality, that there will be an opportunity for the goods to be examined and that the goods will be free from any defects rendering them unmerchantable.

The second situation that is provided for in the Act concerns contracts for the hire of goods. These contracts are not concerned with the transfer of ownership, as in works and materials contracts, or contracts for the sale of goods, but transfer of possession of goods, as when goods are hired for a period of time. Sections 7–10 inclusive are similar in wording and effect to sections 2–5 but are concerned with goods which are the subject of hire contracts.

The second part of the Supply of Goods and Services Act 1982 (as amended by the Sale and Supply of Goods Act 1994) is concerned with the provision of services. A contract for the supply of service may be a contract for a service which also includes the transfer of property e.g. a contract for building work, or may be a contract purely for a service, such as a contract for a hair cut.

Section 13 of the 1982 Act concerns implied terms about care and skill. Section 13 provides that in a contract for the supply of a service where the supplier is acting in the course of a business, there is an implied term that the supplier will carry out the service with reasonable care and skill.

137

Section 14 concerns the implied term about time of performance. Section 14 provides that where under a contract for the supply of a service by a supplier acting in the course of a business, the time for the service to be carried out is not fixed by the contract, there is an implied term that the service will be carried out within a reasonable time.

Section 15 concerns the implied term about consideration. Section 15 provides that where, under a contract for the supply of a service, the consideration for the service is not determined by the contract, there is an implied term that the party contracting with the supplier will pay a reasonable charge.

As can be seen by the wording of these last sections, they do not impose strict liability, but are concerned with reasonableness i.e. the supplier must carry out the service with *reasonable* care and skill. It is lack of care or skill which will make the supplier liable under this part of the Act. This lack of care is akin to negligence, only this time a duty of care does not need to be established as the duty arises through the contract agreed to by the parties.

It can be seen after examining liability under the legislation and common law relating to the sale and supply of goods that the most important measures persons can take when faced with being strictly liable are to ensure that a quality management assurance system includes keeping records of suppliers and other business parties, and ensuring the records are updated regularly. See Chapter 8 for a more in-depth discussion.

6

Fair Trading: Trade Description and Weights and Measures

Outline

The title 'fair trading' could well encompass a number of areas of English legislation, including the Fair Trading Act 1973 itself, and Part 3 of the Consumer Protection Act 1987. The former of the two is addressed towards the end of this chapter, whilst the latter can be found within Chapter 4 in the discussion of product liability. What this chapter aims to deal with are two further areas of legislation, the Weights and Measures Act 1985[1] and the Trade Descriptions Act 1968[2]. Both incur criminal liability, both are enforced by Inspectors from local government Consumer Protection or Trading Standards Departments, and all cases which are brought under these Acts are prosecuted in the Magistrates Court, although defendants have the right to have their case heard before a jury in the Crown Court. Appeals may be heard by the Queen's Bench Division of the High Court, but only on points of law, by way of *case stated*. Being areas which invoke criminal prosecutions, they do not confer any rights to bring civil actions in order to obtain redress. Although as previously mentioned in Chapter 1, criminal courts do have the power to order guilty parties to pay compensation to their victims.

---

[1] Which repeals the Acts of the same name dated 1963 and 1979.
[2] The Trade Descriptions Act 1972 was repealed by Part 3 of the Consumer Protection Act 1987 as was section 11 of the Trade Descriptions Act 1968.

The Weights and Measures Act 1963, as its name suggests, is concerned with quantitative issues, which themselves can obviously influence the perception of quality. The Trade Descriptions Act 1968 on the other hand is very much to do with quality, and as a result, it is this Act which will be examined first.

The Trade Descriptions Act (1968)

Increasingly an area of public concern, mention of the Trade Descriptions Act (TDA) tends to bring to mind dubious practices concerning the motor vehicle trade. Indeed a good deal of case law concerning the Trade Descriptions Act has been generated by that particular area of commerce. Tour operators and holiday companies have similarly contributed to the proliferation of case law concerning trades descriptions. It would therefore be understandable if the initial implications of this particular Act seem to be of concern primarily to quality practitioners within the retail and service industries. However, protecting the general public in such matters is by no means the sole purpose of the Act, which offers a source of protection, not only to domestic consumers, but also to sole traders, partnerships and corporations. What the Act does not set out to do is to enhance in any way contractual obligations, but simply to ensure that customers are not misled or left in a state of ignorance concerning matters of which they should be informed. It should be noted that the Act is only concerned with statements made in the course of a trade or business. It does not cover false descriptions made in the course of genuine private transactions.

Details, Explanations and Implications

Section 1 of the Act states:

> Any person who in the course of a trade or business.
>   (a) applies a false description to any goods; or
>   (b) supplies or offers to supply any goods to which a false trade description is applied; shall, subject to the provisions of the Act, be guilty of an offence.

Section 1(a): 'applies a false description'

With regard to section 1(a), instances where false trades descriptions are applied are straightforward enough, in that such a description would mislead a purchaser and as a result, in the eyes of the law, is worthy of condemnation. As to whether or not the application of a false trades description involves an act of dishonesty is irrelevant. The offences created by s.1 are of strict liability, and as a result it is not necessary for the prosecution to prove any dishonesty or dishonest intent, this point being confirmed in the High Court during the case of *Alec Norman Garages* v. *Phillips* (1985). Given the inexactitude and generalizations which are used in everyday speech and communication, the removal of the necessity for the prosecution to establish dishonesty or dishonest intent, should be a source of considerable concern which, in turn, should give rise to quality procedures designed to apply stringent controls regarding the descriptions applied by any business to its products or services. Unfortunately this is not always the case, a fact which is highlighted in *Baxters (Butchers)* v. *Manley* (1985), a case which itself provides arguably one of the best examples for employing a quality assurance system when dealing with areas covered by the law (see Chapter 2). A further point to consider is that offences under s.1 are not restricted only to sales; a purchaser can also be held liable especially where he or she has expert knowledge with regard to the subject of the transaction. In *Fletcher* v. *Bugden* (1974) a car dealer purchased a vehicle for a token sum claiming that it was only suitable for scrap. The dealer subsequently repaired the vehicle and advertised it for resale. In so doing he was held to be guilty of applying a false trades description – that the vehicle was only fit for scrap – in the course of his trade or business.

Section 1(b): 'supplies or offers to supply'

Section 1(b): 'supplies or offers to supply any goods to which a false trade description is applied'; should be of particular concern to those involved within the retail sector. It creates an offence which is most likely to occur when a retailer displays goods which have a description which was originally applied by the retailer's own supplier, the manufacturer or importer. In this respect the Act may seem to place a harsh and unfair burden upon retailers, a point which was illustrated by the case of *Sherratt* v. *Geralds the American Jewellers Ltd* (1970). In this case it was stated that Geralds had sold a watch with the descriptions of 'diver's watch' and 'waterproof.' When the watch was tested by the purchaser the watch filled with water and ceased to function. Geralds, having not applied the descriptions themselves, were nevertheless held accountable under section 1(b) for supplying goods to which a false description had been applied.

Invitation to treat?

For those familiar which contract law, one possible avenue of defence for the retailer may appear to be offered by the case of *Fisher* v. *Bell* (1961) (see Contract, Chapter 2) where a shopkeeper was acquitted of 'offering to supply' on the grounds that the goods on display were in fact providing an 'invitation to treat.' This seemingly appears to provide the retailer with an all encompassing defence. However, the Trade Descriptions Act specifically prevents this defence by defining the term 'offering to supply' in section 6 of the Act which specifies: 'A person exposing goods for supply or having goods in his possession shall be deemed to offer to supply them.' It was this clause which was to be the downfall of the defence in the case of *Haringey London Borough* v. *Piro Shoes Ltd.* (1976) – a case where corrective action seemed to have been taken. The defendants sold shoes which although marked 'all leather' were not entirely made from leather. Instructions were issued to all the shop managers instructing them to remove the word 'all' from the shoes' description prior to them leaving the shop. The company were initially acquitted; however, upon appeal by the prosecution the High Court held that an offence had been committed at the time when the shoes were actually offered for sale – 'exposed for supply' In this case the company's corrective actions were directed at a point in time – before the goods left the shop – which was too late, an offence under the Trade Descriptions Act having already been committed.

What constitutes a trade description?

A list of what is included in section 2 of the Act, together with some suggested interpretations, are given in Table 6.1.

It should be borne in mind that according to s.2 of the Act; 'a trade description is an indication, direct or indirect, and by what ever means given . . .' It is not necessary for a trade description to be in the form of a written document. This fact is made clear by s.4(2) which states: 'An oral statement may amount to the use of a trade description.' An apparent omission within the Act is the lack of specific reference to pictorial representation. However in the case of *Yugotours* v. *Wadsley* (1968) a holiday brochure depicted a three masted schooner to which the text in the brochure made the reference 'being under full sail on board this majestic schooner'. This in turn led the customers to believe that this was the particular vessel upon which they could expect to be sailing during their holiday. When Yugotours actually provided a twin masted schooner without sails, they were held to be guilty of making a false statement in contravention of the Act.

Table 6.1   TDA section 2(1) and some possible interpretations

Referred to in s.2 as:	*Possible interpretations*
(a) quantity size or gauge	size of clothing, length of material etc. Note s.2(3) specifies that 'quantity includes length, width, height, area, volume, capacity, weight and number'
(b) method of manufacture, production, processing or reconditioning	the term 'hand made'
(c) composition	garment labelled 'pure new wool'
(d) fitness for purpose, strength, performance, behaviour or accuracy	a watch specified as being 'waterproof,' vehicle speeds or fuel consumption;
(e) any physical characteristics not included in the preceding paragraphs	car equipped with four track stereo
(f) testing by any person and the results thereof	tested in accordance with British Standard number . . .
(g) approval by any person or conformity with a type approved by any person	golf clubs as used by Nick Faldo
(h) place or date of manufacture, production, processing or reconditioning	made in Britain, first registered in 1996
(i) person by whom manufactured, produced, processed or reconditioned	produced under licence from Coca Cola
(j) other history including ownership or use	one careful lady owner

So what constitutes a false trade description?

According to s.3(1), 'A false trade description is a trade description which is false to a material degree.' In 1970 the High Court considered the issue of 'false to a material degree' when it heard the prosecution's appeal in the case of *Donnelly* v. *Rowlands* (1970). The case had been brought because Rowlands had been retailing milk in bottles which were embossed 'Express C.W.S, Northern and Goodwins', rather than with his own name, trade mark or logo. However, the bottle caps clearly displayed the words 'Untreated milk. Produced from T.T. cows. Rowlands.' In dismissing the appeal, the court decided that the words on the bottle cap were an accurate trade description of the contents, and that the embossing on the bottle only referred to the bottle itself. As a result the description on the bottle was not 'false to a material degree.'

False or misleading?

The Act does not limit itself to those statements or indications which themselves prove to be false, but also encompasses those which are so misleading that they can be judged to be 'false to a material degree.' In such instances ss.3(1) and 3(2) apply. The following three cases give examples of misleading statements which were judged to constitute false trade descriptions.

The first; *R. v. Inner London Justices* (1983) concerned an appeal brought by the crown in order to try to quash an acquittal. The case concerned a motor vehicle which had been described as having only one owner. In actual fact the vehicle had belonged to a car leasing company, and had during that period of ownership five different keepers. The court's decision was that the term 'one owner' meant that the car had been controlled and maintained by only one person. As a result, to describe the vehicle in this case as having only one owner, was misleading and as a result 'false to a material degree.'

The second case, *Dixon v. Barnett* (1989) centred around a description which, whilst scientifically correct, did in fact prove to be misleading to a customer without scientific knowledge. The company had advertised a telescope as having a magnification of 455 times, which in scientific terms proved to be true, although at that magnification the view it gave was blurred. It transpired that 120 times was in fact its maximum useful magnification. The court held that an ordinary customer would only be interested in the useful magnification of the instrument and as a result the statement was considered to be so misleading as to be false to a material degree. This is an example which should act as cautionary tale not only to the retail sector but also to all manufacturing industries who may well be held accountable for claims made on their packaging.

The third and most recent case concerned a ruling in the High Court concerning the availability of goods advertised at the point of sale. In *Denard v. Smith and another* (1990) it transpired that goods which were offered for sale were not immediately available as they were temporarily out of stock. The court decided that if customers are not informed at the time of purchase as to the non-availability of the goods advertised, then a false trade description has been applied.

The reference to standards

One further point which needs to be highlighted with regard to what constitutes a false description, is made by the Act in its reference to standards. In s.3(4) it says: 'A false indication, or anything likely to be taken as an indication which would be false, that any goods comply with a standard specified or recognized by any person or implied by the approval of any person shall be deemed to be a false trades description, if there is no such standard so specified, recognized or implied.' This section would appear to be solely concerned with the less scrupulous if not downright dishonest. However, the unwary should take note that, whilst this section concerns itself with standards which do not exist, false claims made regarding compliance with existing standards may well be covered by other sections of the Act, notably section 2.

Knowingly false; or reckless?

Section 14 of the Trade Descriptions Act (1968) creates two additional criminal offences which are of direct concern to those operating in the service sector.

Section 14-(1) states:

It shall be an offence for any person in the course of any trade or business:
(a) to make a statement which he knows to be false; or
(b) recklessly to make a statement which is false; as to any of the following matters:
   (i) the provision in the course of any trade or business of any services, accommodation or facilities:
   (ii) the nature of any accommodation or businesses . . .;
   (iii) the time in which, manner in which or persons by whom any services, accommodation or facilities are so provided; or
   (iv) the examination, approval or evaluation by any person of any services, accommodation or facilities so provided; or
   (v) the location or amenities of any accommodation so provided.

Section 14 proves to be the hardest section of the Act under which to gain a conviction. Whereas for convictions to be obtained for contraventions of s.1 of the Act required no necessity on behalf of the prosecution to prove dishonesty or dishonest intent, s.14 does not create offences of strict liability, and as a result places an obligation on the prosecution to prove either:

1 that the defendant knew the statement to be false;
2 that the defendant was reckless in making the statement, in other words that the defendant made the statement irrespective as to whether it was true or false.

In legal terminology the prosecution must show *mens rea* (guilty mind), hence the difficulty in obtaining convictions. Two further points to note are made in s.14(4) which states: 'In this section "false" means false to a material degree and "service" does not include anything done under a contract of service.' The latter point being covered by contract law.

'. . . statement which he knows to be false'

As previously stated, offences under s.14 are the most difficult to gain convictions against. The following case is an example of just how difficult it can be.

The case in question is that of *Sunair Holidays* v. *Dodd* (1970). In this case, the holiday company advertised a package holiday where all the twin bedded rooms were complete with a bath, shower and terrace. Indeed the contract between Sunair and the hotel owners was for accommodation of that description. A case arose when a customer who booked that particular holiday was actually allocated a room without a terrace. At the time that the statement was made (the brochure was printed) Sunair knew that the accommodation existed and had every intention of providing it. The fact that Sunair had not checked with the hotel to ensure that their customers were actually given this accommodation and as a result had no knowledge to the contrary it was adjudged that no offence was committed.

It is at this point extremely important to stipulate that each case has to be judged on its own set of circumstances, and in no way are the authors suggesting that the obvious lack of quality procedures, with regard to vendor appraisal, provides a defence in law. However, contrast this case with that of a similar one, namely *Wings Ltd* v. *Ellis* (1984).

In 1981 Wings published a brochure which described a hotel in Sri Lanka as having air conditioned bedrooms, an inaccurate statement which at the time they believed to be true. Later that same year when Wings discovered the error, they informed their agents, and the customers who had already booked a holiday in that particular hotel. The contravention of the Act occurred when, in the following year, a holiday was booked on the basis that the hotel rooms had air conditioning. In the High Court, Wings argued successfully that no offence had been committed because at the time of publication they were unaware of the error, and that at the time the holiday was booked they believed that the mistake had been rectified. Unfortunately for the defendants, the prosecution were allowed to appeal to the House of Lords which concluded that an offence had been committed when the holiday was booked, because at that time the holiday company knew that the statement regarding air conditioning was false. The fact that Wings were unaware of the falsity of their statement when the brochure was first published was considered to be irrelevant.

It would be quite wrong to infer from these two cases that ignorance is a form of defence against legal proceedings. Nevertheless it does seem harsh that the company which attempted to rectify their error – implemented corrective action – were held accountable by law, whilst the company which did not, were, with regard to the Trade Descriptions Act, held to be blameless. The rider – with regard to the Trade Descriptions Act – is included because it is extremely difficult, if not impossible, to quantify the effects which such legal proceedings have on subsequent trade. It would seem reasonable to suggest, that regardless of the old adage, 'there is no such thing as bad publicity,' litigation of this kind can hardly be good for business. The lesson to be learned from the experience of both cases must surely be the necessity to ensure the verification of purchased products and conduct contract review procedures, prior to the acceptance of any order.

Reckless statements

The following is not offered as an illustration as to where quality procedures could have been applied, but rather as an indication of what can be considered to be a reckless statement, and as a caution, as to the care which is required to be taken in this area.

In 1973 a prosecution was brought against MFI Warehouses regarding an advertisement which had been published only after it had been considered by the company's chairman. The advertisement offered louvre doors for sale by mail order, on fourteen days free approval, at the end of which the price plus carriage was to be paid. In the same advertisement folding door gear was advertised also on free approval for fourteen days 'carriage free'. It was intended that the two items should be sold together, and on that basis carriage would not be charged on the folding door gear. The defendants' chairman did not appreciate the ambiguity of the advertisement. As a result, when a customer ordered only the folding door gear he was also charged for the carriage. This resulted in conviction, under s.14 of the Trade Descriptions Act, which was upheld on appeal to the High Court.

Multiple prosecutions

Autrefois convict is a long standing legal rule which means that a person cannot be charged more than once for an offence for which they have already been found guilty, or for which they have been acquitted. As regards the Trade Descriptions Act this is not necessarily a defence which will be accepted. In the case of *R*. v. *Thomson Holidays* (1974) the company were convicted under s.14 of the Trade Descriptions Act, in a prosecution brought by one local authority, for publishing a brochure containing a false statement. When a second prosecution, for the same offence, was instigated by a different local authority the company pleaded *autrefois convict* with regard to the previous conviction. The Court of Appeal held that a false statement was made by the company every time that the brochure was read by someone, and as a result a fresh offence was committed each time. On this basis it is theoretically possible for a company to be prosecuted on numerous occasions for a single false statement which has been committed to print. A sobering thought for every company which advertises its products or services.

Penalties

As previously mentioned, offences committed contrary to the Trade Descriptions Act 1968 are dealt with in the criminal courts. Section 18 provides for two penalties:

> 18(a) 'on summary conviction [that is in the Magistrates Court] a fine not exceeding £2000';
> 18(b) 'on conviction on indictment a fine [unlimited] or imprisonment not exceeding two years or both'.

In cases where conviction is obtained on indictment the fine imposed is unlimited by legislation and therefore subject to the decision of the court.

The sum of £2000 mentioned above was a revised figure set in 1984. The original maximum fine was £400. It should also be remembered that multiple prosecutions for the same offence are a possibility – *R*. v. *Thomson Holidays Ltd* (1974) – and with that comes the additional possibility of a separate fine being imposed for each prosecution. Whilst terms of imprisonment are an option under s.18(b), the High Court, in *R*. v. *Haester* (1973), stated that imprisonment was normally reserved for cases involving dishonesty. As a general rule it would appear to be dishonest vehicle dealers who seem to attract both heavy fines and prison sentences.

Civil actions and compensation

The Trade Descriptions Act does not provide for civil redress. However, a criminal conviction may assist a consumer seeking compensation which can be awarded by the courts under the Powers of Criminal Courts Act (1973). In addition, by virtue of the Civil Evidence Act 1968, a criminal conviction under the Trade Descriptions Act may be used as evidence in a civil action for compensation.

Defences

Section 24 of the Trade Descriptions Act provides for a number of defences, which places the onus on the defendant to prove:

> 24–1(a) 'that the commission of the offence was due to a mistake or to reliance on information supplied to him or to the act of default of another person, an accident or some other cause beyond his control; and
> 24–1(b) 'that he took all reasonable precautions and exercised due diligence to avoid the commission of such an offence by himself or any other person under his control.'

In order for a defence to be successful, the defendant must prove the existence of at least one of the reasons given in s.24–1(a), together with all of those expressed in s.24–1(b).

It is in attempting to prove these defences that a good quality assurance system may prove to be of great assistance. The following two cases are offered as examples of where good quality assurance practices and the application of suitable procedures could well have been of worth to the defendants in each case.

Firstly, in *Denard* v. *Smith* (1990) the defendant was accused of falsely advertising goods which were out of stock. The defendant sought the defence offered by s.24. However, the court held the defendant would have been better placed to prove 'due diligence' if some form of instruction had been issued with regard to the alteration of literature displayed at the point of sale. A notice attached to the main advertisement would have been sufficient. In essence it was the absence of work instructions or working procedures, a fundamental part of any quality assurance system, which proved to be the defendant's downfall in this particular case.

The second case, of *Baxters (Butchers)* v. *Manley* (1985) highlights several areas where a quality management system could well have assisted the defence, or alternatively prevented the offence in the first place. The case brought against Baxters involved breaches of both the Trade Descriptions Act 1968 and the Weights and Measures Act of 1963. The case concerned the misrepresentation of the weight of meat exposed for sale, and exposing meat for sale ostensibly at a lower price than the customer was actually charged. In court Baxters admitted that breaches of the Trade Descriptions Act and the Weights and Measures Act had occurred, but claimed that this was a result of the acts or by default of the manager of the particular shop involved. This claim was accepted. However, what the Court did not accept was that Baxters had taken all reasonable precautions to avoid the offence taking place, nor had they exercised all due diligence. It was the Court's decision that the statutory defences under s.24 of the Trade Descriptions Act 1968 and s.26 of the Weights and Measures Act 1963 were not, on the balance of probability, made out because:

(a) Baxters failed to establish that any positive precautions were taken to prevent the commission of the offence;
(b) there was inadequate checking and supervision by the district manager;
(c) no detailed instructions or guidelines were given to the manager on how to comply with the relevant statutory provisions;
(d) there was insufficient staff training.

Given leave to appeal on the grounds of *case stated*, it was held that: 'it could not be said that there was insufficient evidence to justify the justices' conclusions.' In giving its decision, the court was not setting out any standard of what was required. It is in each case for the justices to decide on the evidence presented to them as to whether or not the statutory defences are made out. In defending the case, Baxters relied on that part of section 24-(1)(a) which states: 'that the commission of the offence was due to . . . the act or default of another person . . .' In this case, the 'person' was not an outside supplier, but the manager of the particular shop involved. In order to use this defence, the defendants themselves must inform the prosecution, at least seven days in advance of the hearing – section 24-(2) of the Act – as to whom they believe the actual person to be. This in turn, would allow the prosecution to proceed against that person. In this particular case it was accepted that the offences were the result of the 'act or default' of the shop manager, but fortunately for him and unfortunately for Baxters their defence failed because of the reasons mentioned above.

Examined from the point of view of a quality practitioner, had Baxters employed a quality system, such as that which would have qualified for approval to the BS EN ISO 9000 series, and, had that system been fully operational and adhered to, it should be clear how these legal proceedings may have been avoided. Indeed had such a quality system been in place a quality audit would have shown up each of the points (a) to (d), highlighted by the court case, and marked them as a source of non conformance.

Legal consequences for employees

An additional point illustrated by this case, is the need for quality procedures not only to be in place, but also the need for them to be correctly adhered to. Failure to do so can result in *legal consequences for employees of a company* whilst providing a defence for the company itself. This principle was established by the House of Lords when they heard the appeal in the case of *Tesco Supermarkets* v. *Nattrass* (1972). This involved packs of washing powder being advertised at a special offer price. However, at the time a purchase was made, the special offer packs were out of stock, with the result that only packs sold at the full price were on display. The prosecution was

Table 6.2   Court decisions: *Tesco* v. *Natrass* (1972)

Court	The court's decision
Magistrates (initial prosecution)	The defendant had taken all reasonable precautions as required by s.24(1)(b) but; the store manager *did not* constitute 'another person' under s.24(1)(a) as a result the defendant (Tesco Supermarkets) was convicted
High Court (appeal against conviction)	The store manager *did* constitute 'another person' under s.24(1)(a) but; as the store manager was one of those responsible for operating the system of supervision and had failed to do so, the defendant could not claim to have taken all reasonable precautions as is required by s.24(1)(b). As a result the appeal was dismissed
House of Lords (appeal on a point of law which is of general public interest)	The court held that the company had in fact set up a proper system of control and as a result had complied with s.24(1)(b). The court further held that the defendant were those individuals who were '... in actual control of the operations ...' or those individuals who are '... not responsible to another person in the company for the manner in which he discharges his duties ... The store manager was not considered to be within the scope of this ruling and as a result constituted 'another person' for the purposes of s.24(1)(a) As a result the conviction was quashed

brought under the now repealed section 11 (see Consumer Protection Act 1984 Part 3). The defence claimed that the store operated a comprehensive system of management to ensure that offences against the Trade Descriptions Act did not occur. The defence further claimed that the offence occurred due to the actions of a store manager who had failed to check the work of his staff. The case was tried, and subsequently on appeal went all the way to the House of Lords. The decision of each court is shown in Table 6.2.

The table indicates not only the various ways in which the Act was interpreted with regard to this case, but more importantly it set two legal precedents; firstly those who persons who could be held responsible for the company's actions, and secondly that direct employees of the company can be considered to be 'another person' and as a result be liable under the Act.

Additional defences

Two additional forms of defence are provided for within the Trade Descriptions Act. Section 24-(3) states:

> 'In any proceedings for an offence under this Act of supplying or offering to supply goods to which a false trade description is applied it shall be a defence for the person charged to prove [emphasis added] that he did not know, and could not with reasonable diligence have ascertained, that the goods did not conform to the description or that the description had been applied to the goods.'

Note that in this section the burden of proof is placed upon the defendant rather than the prosecution.

The other defence is contained in section 25 of the Act and concerns the publication of advertisements.

> 'In the proceedings for an offence under this Act committed by the publication of an advertisement it shall be a defence for the person charged to prove that he is a person whose business it is to publish or arrange for publication of advertisements and that he received the advertisement for publication in the ordinary course of business and did not know and had no reason to suspect that its publication would amount to an offence under this Act.'

This form of defence which again places the burden of proof upon the defendant protects not only the publishers of newspapers and periodicals but also advertisement agencies and those who arrange for the publication of advertisements.

Disclaimers

Though not specified as part of the Trade Descriptions Act, disclaimers can provide a source of defence in certain circumstances, namely those covered by section 1(b) 'supplies or offers to supply goods to which a false trades description has been applied.'

However, for the use of disclaimers to be successful a number of criteria have to be fulfilled. According to the ruling given by the then Lord Chief Justice in the case of *Norman* v. *Bennet* (1974), a disclaimer must be 'as bold, precise and compelling as the trade description itself and must be as effectively brought to the notice of any person to whom the goods may be supplied.' He continued by saying '. . . in other words the disclaimer must be equal to the trade description in the extent to which it is likely to get home to anyone interested in receiving the goods.' This ruling presents any disclaimer with two test of its effectiveness:

1 that it should be 'bold, precise and compelling';
2 the time at which is applied.

For the benefit of the motor trade at least, the first point was resolved in 1976, when the Director General of Fair Trading approved the motor trade Code of Practice which laid down a specific form of wording to be used when referring a vehicle's recorded mileage.

As regard the second point, bearing in mind the Act states: 'or offers to supply . . .', this point was reinforced in the case of *Doble* v. *David Greig Ltd* (1972). The defendants offered bottles of blackcurrant juice for sale with the indication that a refundable deposit was payable upon the return of the bottle. A second notice at the point of sale contradicted this statement. The prosecution was upheld on the basis that the offence of offering to supply was committed when the goods were put on display. Clearly there would be little to be gained in attempting to disclaim liability once an offence has been committed.

TDA and the Quality Issues

Throughout this section, individual legal proceedings have been used not only to try to explain the nature of the Trade Descriptions Act 1968, but also to demonstrate how the Act has been applied, and to indicate areas in which the establishment and use of a quality assurance management system may have been of use in preventing the prosecutions which have taken place. Earlier in the chapter, little was made of the corrective action taken by Spiro Shoes in

Table 6.3   Legal proceedings and relevant areas of QAMS

Legal proceedings	Subject	Relevant areas of QAMS
Sherratt v. *Geralds the American Jewellers Ltd* (1970) (see p. 141)	Waterproof watch – goods to which a false description is applied	Verifying the product at the jewellers suppliers, or inspection and testing of the product upon receipt may have prevented the litigation. Alternatively the existence and implementation of such procedures may have assisted the defence – '. . . took all reasonable precautions . . .', '. . . exercised all due diligence'
Wings v. *Ellis* (1984) (see p. 146)	Holiday accommodation	Again verification of the product – the accommodation – should have taken place
MFI v. *Nattrass* (1973) (see p. 147)	Louvre doors	Whilst seemingly arising from the error of an individual, a document or data approval procedure, possibly by 'in house council or legal executive' should have prevented the situation arising
Baxters (Butchers) v. *Manley* (1985) (see p. 149)	Sale of meat	Demonstrated the need for procedures covering: 1 the implementation of work instructions (document procedures) which themselves govern the product or service 2 preventive actions 3 training

the case of *Haringey London Borough* v. *Spiro Shoes* (1974). Similarly Table 6.3 makes no reference to the corrective action taken by Wings Ltd in trying to prevent the prosecution which subsequently came about.

The reason for these omissions is not to suggest that corrective action procedures should themselves be omitted, but because the two cases in question have been decisive in indicating the worth of such procedures with regard to this particular area of legislation. As was shown in both cases, the use of corrective actions, if applied after the goods are displayed, do not prevent prosecution. Nor in fact does their presence constitute either of the defences of 'due diligence' or 'all reasonable care.' The lesson to be learned with regard to corrective action procedures, and the Trade Descriptions Act 1968, is that the procedures need to be part of a quality system which acknowledges that errors will occur, but sets out to eliminate the potential effects of such errors prior to any goods or services being offered to potential customers.

One other area where quality assurance procedures can either prevent a situation which leads to a contravention of the Trade Descriptions Act, or alternatively may aid with a legal defence, concerns the subject of sampling inspection. In the United Kingdom statistical sampling procedures are covered by the British Standard BS 6000 series. The implementation of the sampling plans specified within these standards is designed to give the concerned parties a degree of confidence that any samples taken are representative of what is actually being produced. However, it is accepted that their are risks involved to both the consumer and the producer, because a good sample is not necessarily indicative of a good batch, and similarly a bad sample is not necessarily indicative of a bad batch. However, failure to conduct this type of sampling or some form of sampling which attempts to afford the same degree of confidence in the product being supplied, could well result in the inability to establish either of the defences of 'reasonable precautions' or 'due diligence' which are acceptable under the Trade Descriptions Act. This proved to be the situation which arose in 1988 in the case of *Rotherham Metropolitan Borough Council* v. *Raysun (UK.)* The prosecution which was brought concerned the presence of toxic material in wax crayons which were described as 'poisonless.' The company, which imported the crayons from Hong Kong, employed a system whereby their agents in Hong Kong were required to have samples of the crayons analysed, but were only required to report on any adverse results. In England, the defendants having not received any adverse reports, had tests of their own conducted on one packet from 120,000. In the absence of any information the High Court decided that such a sample '. . . did not indicate the standard of care required by the statutory provisions.'

The Weights and Measures Act 1985

The Weights and Measures Act, as the name suggests, deals with quantitative issues, and like the Consumer Protection Act, imposes only criminal liability, with the task of enforcement being the responsibility of local government through the offices of their local Consumer Protection or Trading Standards departments. Within the Act, units of measurement, both metric and imperial, are defined for: length, area, volume, capacity and mass or weight, and having specified these units the Act declares it to be an offence to use units other than those specified. However, under the terms of the Act, the Secretary of State may make an order which permits changes to the units specified, or the ways in which they are applied. Examples of this are that from 1 October 1995 all pre-packed foods have been labelled in the metric units of: kilograms and grams, and from 1 January 2000 all foods sold loose will also have to be sold using these metric units. In both instances trades will retain the right to use imperial units but only in addition to the metric measures.

'Use for trade'

In determining what is covered by the Act Part II specifies the following:

1 that 'the transaction is by reference to quantity;' and
2 that the transaction is for 'the transferring of money or money's worth.'

The Act makes it clear that it only applies to sale by retail, and that any sale must include the transferring of money or money's worth. The Act does not apply to goods to be despatched to countries outside Great Britain unless specifically designated otherwise.

In determining specific responsibility section 7(5) states:

'Where any weighing or measuring equipment is found in the possession of any person carrying on trade or on any premises which are used for trade, that person or as the case may be, the occupier of those premises shall be deemed for the purposes of this Act, unless the contrary is proved, to have that equipment in his possession for trade.'

In examining this section it is important to notice the phrase 'unless to the contrary is proved'. As with the Trade Descriptions Act, the Weights and Measures Act puts the onus on the defendant to establish an acceptable defence. The requirement on the prosecution is simply to establish that an offence has been committed.

General Offences

Within the Act there are in excess twenty sections or subsections which relate to possible offences, for which the penalties range from the imposition of fines to terms of imprisonment or both. A number of these offences are specific to certain areas, such as the use of the carat (metric) being used specifically for the purposes of transactions in precious stones or jewels, and as a result are beyond the scope of this publication. However, the Act also includes, within Part IV, a number of general offences, a summary of which is given in Table 6.4.

Table 6.4a Weights and Measures Act 1963: Part IV: Summary of offences

Offence	*Area concerned*
Section 22 Offences in transactions in particular goods	Goods specified as: Foods stuffs (schedule 4) Sand and ballast (schedule 5) Solid fuel (schedule 6) Liquid fuel (schedule 7) Composite goods and collections of articles (schedule 8)

Section 23
Quantity to be stated in writing
in certain cases
	This concerns the sale of goods which
	according to the Act:
	(a) need to be sold 'by quantity expressed in a
	    particular manner'; and
	(b) when 'sold expressed in a particular
	    manner' as required by the Act, this fact
	    needs 'to be made known to the buyer at or
	    before a particular time'

Section 24 Short weight	Concerns the provision or the misrepresentation of any short quantity (not only weight)

Table 6.4b Weights and Measures Act 1985: summary of offences

Offence	*Area concerned*
Section 28 Short weight etc.	Shortages of goods being sold by weight, other measurement or by number
Section 29 Misrepresentation	Misrepresentation may be written, oral or '. . . any other act calculated to mislead a person buying or selling . . .
Section 30 Quantity less than stated	Concerns the marking of goods or their containers with information which subsequently proves to be incorrect (see defences – subsequent deficiencies)
Section 31 Incorrect statements	Refers to any documentation which is associated the goods being sold and which contains information is 'materially incorrect'
Section 32 Offences due the default of a third person	Very similar to s.24(1)(a) of the Trade Descriptions Act – 'act or default of another person'

Certified Equipment

In addition to the offences summarized in Table 6.4, the Act concerns itself in some detail with the design (pattern) and construction of all weighing and measuring equipment which is used for trade. It also specifies that all such equipment must, prior to its use, be checked for accuracy by a qualified inspector, and be stamped to indicate that it has passed inspection. This provides a permanent record of the equipment's initial calibration and fitness for use. However, such an inspection does not withdraw the obligation to maintain the equipment within the specified limits. As a result quality procedures should be instigated to ensure regular calibration is undertaken. Failure to maintain the accuracy of measuring and weighing equipment could well result in criminal prosecution.

Defences

Warranty

A number of defences are provided for under the Weights and Measures Act, the first of which. 'pleading warranty as a defence' highlights the need for certificates of conformance or compliance. Properly designed, two of the features of such a certificate would be a description of the product, and the quantity being supplied. In attempting to use warranty as a defence it would be necessary to:

1 produce a certificate of conformance, depicting the quantities in question;
2 show that at the time of the offence, there was no reason to believe that the quantity was other than that stated on the warranty (certificate of conformance);
3 also show that where 'the warranty was given by a person who, at the time he gave it was resident outside Great Britain and any designated country, that the person charged had taken all reasonable steps to check the accuracy of the statement';
4 demonstrate that all reasonable steps had been taken to prevent any change in the quantity of goods whilst they had been in the defendant's possession; and
5 prove that at the time of the offence, apart from the quantity, the goods were in the same state as when purchased.

As mentioned previously, to issue a false warranty as to the quantity is itself an offence under the Weights and Measures Act (1985). A further point to be considered is that to issue such a warranty may also result in prosecution being brought under the Trade Descriptions Act 1968, as was the case in *Baxters (Butchers)* v. *Manley* (1985).

Reasonable precautions and due diligence

Section 34(1) states:

> '. . . it shall be a defence for the person charged to prove that he took all reasonable precautions and exercised all due diligence to avoid the commission of such an offence.'

It was this defence, which was at the time included in the Weights and Measures Act (1963), which was used in the case of *Baxters (Butchers)* v. *Manley* and, as previously discussed, it highlights an area of litigation where the implementation of good quality procedures could well have either prevented the prosecution arising in the first instance or failing that, could have helped to provided Baxters which a sustainable defence to the charges brought.

Subsequent deficiency

Section 35 provides a defence for situations where the quantity of goods:
(a) is less than that indicated on the container;
(b) is less than that indicated on documentation which is associated with the goods;
(c) which are required to be pre-packed or supplied in specific quantities is less than indicated.

The Act provides that, '. . . it shall be a defence for the person charged to prove that the deficiency arose:'
(i)   in the case of (a) '. . . after the making up of the goods and the marking of the container;'
(ii)   in the case of (b) '. . . after the completion of the documentation;'
(iii) in the case of (c) '. . . after the making up or the making, as the case may be, of the goods for sale;

'and was attributable wholly to factors for which reasonable allowance was made in stating the quantity of the goods in marking or document or in making up or making the goods for sale, as the case may be.'

The two points to be noted here are:

1 the terms 'attributable wholly,' and 'reasonable allowance;' and
2 that once again the burden of proof is placed upon the defendant, the prosecution merely being required to prove that the offence has been committed.

Weights and Measures and the Quality Issues

As most quality practitioners will be aware, it is documentation, in terms of its absence or lack of completion, which forms the most prolific single source of non-compliance during the auditing of quality assurance management systems. An examination of the requirements of the Weights and Measures Act suggests a similar importance needs to be placed upon documentation if contraventions of the Act are to be avoided, or if a successful defence is to be established. Particular areas of a quality system, other than documentation control, which will need to be addressed are:

- the control of inspection measuring and test equipment;
- sampling; and
- training.

Additionally, bearing in mind the outcome of *Baxters* v. *Manley,* procedures will need to be devised for the relevant areas of a quality system which are mentioned in Table 6.3.

It would appear to be trite to suggest that procedures which are written should aim at ensuring that the product or service conforms to the customer's requirements. That after all is the intention of standards such as BS EN ISO 9000. What is not so obvious is that quality procedures concerning the areas mentioned in this chapter should be written with the Trade Descriptions Act 1968 and the Weights and Measures Act 1985 uppermost in mind. Whilst there are defences provided for under these Acts, such as 'due diligence' (Weights and Measures), for which the existence and operation of quality assurance procedures *may* play a part, there is no room within either piece of legislation for the type or scope of concessions which may be granted by a commercial customer. Given the punitive nature of both pieces of legislation and, in the case of Weights and Measures, the relative ease of proving an offence has been committed, the emphasis must be placed on right first time.

The Fair Trading Act 1973

The Fair Trading Act of 1973 created the post of Director of Fair Trading and with it the Office of Fair Trading itself. The Act confers a number of powers on the Director, as does the subsequent Consumer Credit Act 1974 and the Estate Agents Act 1979. In brief the powers of the Director of Fair Trading are to:

1 review commercial practices;
2 make referrals concerning adverse trade practices to the Consumer Protection Advisory Committee;
3 bring actions against traders whose practices are persistently unfair to consumers;
4 supervise the enforcement of the Consumer Credit Act (1974);
5 arrange the publication of advice and information for consumers;
6 encourage the publication of codes of practice by trade associations;
7 oversee the working of the Estate Agents Act (1979).

Whilst the Director of Fair Trading can bring proceedings, it is likely that such actions will be as a result of contraventions of the Trade Descriptions Act (1968) and the Weights and Measures Act (1985), both of which, as can be seen, provide a need for well constructed and correctly implemented quality procedures to prevent situations which could result in legal proceedings being taken.

7

Food Safety Act 1990 and General Food Hygiene Regulations 1995

Of all industries referred to in this book the food industry is the one which ought to be totally dedicated to the maxim of quality. As with any other industry, there will always be the variety between organizations with prestigious names and those with cut price labels, but irrespective of whether the product is baked beans or caviare, from a producer or processor, wholesaler or retailer, restaurateur or cafe owner, all have one basic thing in common – the need to provide food which is fit for human consumption – that which is fit for its purpose, in other words that which is a quality product.

Outline of the Act

The Food Safety Act 1990 came into effect on 1 January 1991. In so doing, it repealed seven former Acts of Parliament completely, together with sections of a further fourteen Acts and, as a result, it strengthened and updated the law on both food safety and consumer protection, with regard to food, in the United Kingdom. Irrespective of the size of the business, the Act concerns everyone working in the sale, distribution, production, processing or storage of food, including the self employed and non-profit making organizations.

The Act was created with five basic aims in mind:

1 to ensure that food which is produced for sale is in fact safe to eat;
2 to ensure that food is not misleadingly presented;

3 to strengthen further the legal powers and penalties which already existed;

4 to ensure that the United Kingdom could fulfil its role in the European Union; and

5 to maintain pace with environmental change.

Like the laws relating to consumer protection, the Food Safety Act 1990 is enforced by both local as well as central government, with the day to day enforcement being the responsibility of the local authorities. The need for central government to become involved may arise in certain emergencies, and the Act does provide the Government with the power to make emergency control orders. In addition, veterinary surgeons appointed by the state may also be involved in enforcement on farms and in slaughterhouses. The main role of central government is, however, in the formulation of food policy. The role of the local authority is the enforcement of the Act in two main areas:

1 with regard to the labelling, and composition of food, which will normally be the concern of trading standards officers; and

2 concerning hygiene and the contamination of food which renders it unfit for human consumption. This area is usually the concern of the local authority's environmental health officer.

Contents of the Act

Divided into four main parts, the Act starts by defining the actual meaning of the word 'food':

'Food': Its Definition under the Act

According to the Act 'food' includes:
  (a) drink;
  (b) articles and substances of no nutritional value which are used for human consumption;
  (c) chewing gum and other substances of the like nature and use; and
  (d) articles and substances used as ingredients in the preparation of food or anything falling within this subsection.

Embraced within these definitions are:

- anything which may be used as an ingredient;
- animals such as oysters which are eaten live;

- any form of drink;
- water which is used in the production of food or drawn from a tap during the course of a food business; and, for the first time:
- items such as dietary supplements and slimming aids.

In addition the Act also covers what it refers to as 'food sources' and defines these as: 'growing crops or live animals, birds or fish from which food is intended to be derived (whether by harvesting, slaughtering, milking, collecting eggs or otherwise)'. As a result, farmers and food producers are now directly involved considerably more than they have been in the past.

A further inclusion in the Act is that of 'contact materials'. These are defined as being:

> 'Any article or substance which is intended to come into contact with food'. This can include anything from wrapping materials to manufacturing equipment. The Act, however, does not include an extended list of specific items under the heading of contact materials. What it does do is provide the Government with the necessary powers to make regulations covering specific details.

Items which are not considered to be 'food' for the purposes of the Act are:

- fodder or feeding stuffs for animals, birds or fish;
- controlled drugs; or
- medicines.

Organizations Covered by the Act

The Act's definition by itself gives it extremely extensive scope. Add to this the Act's further definition of what constitutes a business and the scope becomes enormous. Part 1 Section 1(3) of the Act states that where food is concerned, the term business includes:

> 'the undertaking of canteen, club, school, hospital or institution, whether carried out for profit or not, and any undertaking carried out by a public or local authority.'

It further goes on to state that:

> 'commercial operation, in relation to food or contact material, means any of the following, namely:

(a) selling, possessing for sale and offering, exposing or advertising for sale;

(b) consigning, delivering or serving by way of sale;

(c) preparing for sale or presenting, labelling or wrapping for the purpose of sale;

(d) storing or transporting for sale;

(e) importing and exporting.'

With all of this comes what appears to be a considerable burden for the quality practitioner, for the quality issues are as inextricable from the requirements of the Act as are its requirements for safety. Yet in many respects there is no special requirement for quality procedures which are concerned solely with ensuring that the requirements of the Act are fulfilled. Take for example the inclusion of storage and transportation. The employment of a quality assurance management system in accordance with BS EN ISO 9002: 1994, will necessitate the presence of procedures to ensure the prevention of damage or deterioration of the product during handling and storage. If such procedures are properly constructed and implemented they should automatically ensure that the requirements of the Food Safety Act 1990 are adhered to.

'Extended Meaning of Sale'

As already pointed out above, the term business is not confined to organizations attempting to make a profit. In a similar vein the Act also covers food in situations which would not normally be considered to constitute a sale. Under the heading 'Extended meaning of sale etc.' Part 1 Section 2(2) of the Act states:

'This Act shall apply:

(a) in relation to any food which is offered as a prize or reward or given away in connection with any entertainment to which the public are admitted whether on payment or not . . .'; and

(b) in relation to any food which, for the purposes of advertisement or in furtherance of trade or business, is offered as a prize or reward . . .'

For clarification, the Act states that entertainment for the purposes of this Act 'includes any social gathering, amusement, exhibition, performance, game sport or trial of skill.'

165

The Main Offences

The Food Safety Act 1990 stipulates a number of offences, each of which constitutes a criminal act and may be punishable by the imposition of fines, terms of imprisonment or both. Within the main provisions of Part 2 of the Act, the offences are separated into two discrete groups, namely:

(i) food safety, which in turn includes;
   (a) rendering food injurious to health (section 7); and
   (b) selling food not complying with food safety requirements (section 8);
(ii) consumer protection, which details the offences of:
   (c) selling food not of the nature or substance or quality demanded (section 14); and
   (d) falsely describing or presenting (food section 15).

Section 7 makes it quite clear that a person shall be guilty of an offence if they render food injurious to human health by:

(a) adding any article or substance to the food;
(b) using any article or substance as an ingredient in the preparation of food;
(c) abstracting any constituent from the food; and
(d) subjecting the food to any other process or treatment.

For food to be considered to be injurious to health it would need to be shown that the food in question would be harmful to a substantial proportion of the community, but not necessarily everybody who consumed it. Hence a product which had an adverse effect on children (who obviously constitute a substantial portion of the population) but not necessarily adults, might well be considered injurious to health. Conversely an ingredient which produced an adverse reaction in only a few individuals would probably not be covered.

Section 8 'Selling food which does not comply with food safety requirements' is similarly explicit as to who shall be guilty of an offence when it states that:

'Any person who
(a) sells for human consumption or offers, exposes or advertises for sale ... or has in his possession for the purpose of such sale or of preparation for such sale; or

(b) deposits with, or consigns to any other person for the purposes of such sale or preparation for such sale,

any food which fails to comply with food safety requirements.'

As far as this part of the Act is concerned, food fails to comply with the food safety requirements if:

(a) it has been rendered injurious to health;
(b) it is unfit for human consumption; and
(c) is so contaminated (whether by extraneous matter or otherwise) that it would not be reasonable to expect it to be used for human consumption.

For food to be considered to be unfit it would need to be putrid or toxic, have come into contact with, or contain for example, a dead mouse – the possibilities of such occurrences being a constant source of worry to all involved in the food industry. In addition, as specified in section 8(4)(a) meat 'which had been slaughtered in a knackers yard, or of which the carcass had been brought into a knacker's yard . . . shall be deemed to be unfit for human consumption.'

Additional Offences

The Food Safety Act 1990 also provides for two additional offences, namely:

1 the obstruction of enforcement officers;
2 'offences by bodies corporate'.

In the case of the latter, where an offence is deemed to have been committed by a business or organization (body corporate) directors, managers, company secretaries and the like can also be held responsible for that offence and if found guilty can be punished accordingly.

Food for Human Consumption

It needs to be borne in mind whilst considering the Food Safety Act 1990 that the offences created under the Act are specifically related to food which is for human consumption only. As a result, the Act makes certain assumptions as to what actually constitutes food for human consumption (Section 3).

The Act (Section 3) places the onus of responsibility directly with any defendant to prove to the contrary that, food which can be considered to be commonly used for human consumption, whether it is either sold or offered for sale, or even exposed or kept for sale, is not actually intended for sale for human consumption. The Act further presumes that:

(a) any food which is commonly used for human consumption which is found on premises used for the preparation, storage or sale of that food; and

(b) any article or substance commonly used in the manufacture of food for human consumption which is found on premises used for the preparation, storage or sale of that food'

shall be presumed, until the contrary is proved to be intended for sale, or for manufacturing food for sale, for human consumption.'

Section 3 goes on to point out that:

'any article or substance capable of being used in the composition or preparation of any food which is commonly used for human consumption which is found on premises on which that food is prepared shall, until the contrary is proved, be presumed to be intended for such use.'

Once again it is worth noting where the legal responsibility lies. The presumption of innocence, one of the fundamental principles of English law, is denied to any defendant being prosecuted under this Act. The food industry requires working procedures which are correctly constructed, constantly reviewed and strictly adhered to, in order to assist in achieving a quality product or service. Such procedures being in effect an obvious way of ensuring that the legal requirements, as laid down by the Food Safety Act, are complied with. Further, the industry also has the need for good record keeping and documentation which if properly formulated could well assist in providing the necessary legal defence in the event of a prosecution arising. The food industry is definitely one in which the requirements of a good quality assurance management system can be shown to walk hand in hand with the requirements of the law.

Consumer Protection

Two of the main offences under the Act are concerned with consumer protection, and provide a degree of protection for the consumer in the area of food which is not provided by the Consumer Protection Act 1987 itself. Section 14 makes it an offence for any person to sell '. . . to the purchaser's prejudice any food which is not of the nature or substance or quality demanded by the purchaser . . .' This covers situations such as:

- when the quality or composition are not in accordance with a statutory standard;
- where imitations are passed off as the real article;
- if particular ingredients are below the fair minimum content.

The other piece of consumer protection contained within the Act (section 15) concerns food which is falsely described or presented. This section is in essence to do with labelling and advertising. The Act makes it clear that it is a criminal offence, irrespective of the type of labelling, whether it is '. . . attached to or printed on the wrapper or container . . .', to falsely describe or to describe the contents in such a manner that the purchaser is likely to be misled as to the nature or the substance of the contents.

Who is the Purchaser?

At this point it is worth noting who exactly constitutes a purchaser. In the case of this Act a purchaser can be an individual customer in, for example a shop or restaurant, or one organization buying food from another. It is also worth noting that in keeping with the Consumer Protection Act 1987 the requirements of section 14 apply even if the food is bought for someone other than the purchaser.

Defences

Part 2 of the Food Act provides for the forms of legal defence:

1 offence due to the default of another person (section 20);
2 defence of due diligence (section 21); and
3 defence of publication in the course of a business (section 22).

Offences Due to the Default of Another

Whilst this defence is specifically cited in s.20 and is a defence in its own right, it also forms an important part of the defence of due diligence which is detailed below.

The Defence of Due Diligence

This provides the primary source of defence for proceedings brought under the Food Safety Act 1990. Once again the onus of responsibility is placed upon the defendant to prove innocence in that, in accordance with the requirement of the Act, '. . . he took all reasonable precautions and exercised all due diligence to avoid the commission of the offence by himself or by a person under his control'. In making such a defence it is not necessary for the defendant to provide proof beyond all reasonable doubt. What the defence will have to show is that their case is made out on the balance of probabilities. What constitutes *reasonable precautions* is a matter for the court which, will have to consider the facts of the case as they are presented. In the main, defendants can make a defence on the grounds of due diligence if they can demonstrate that:

(a) . . . the defence was due to the act or default of another person . . . or to the reliance on information supplied by such a person;

(b) he [the defendant] carried out such checks of the food in question as were reasonable in the circumstances, or that it was reasonable for him to rely on checks carried out by the person who supplied the food . . .;

(c) he did not know and had no reason to suspect at the time of the . . . alleged offence that his act or omission would amount to an offence . . .'

In order to be successful with (a) and (b) the defendant would also have to demonstrate the other person in question was not under his or her direct control. As a result such a defence could not be mounted, for example, by a manager or supervisor who cited members of their own department. In addition, to successfully mount a defence which is based upon the default of another, the defendant is required to co-operate with the prosecution by:

(a) serving notice of such a defence at least seven days before the start of the hearing; and

(b) where the defendant has previously appeared before a court in connection with the alleged offence, within one month of his first appearance.

In such instances the defendant will also need to further co-operate with the prosecution by assisting with the identification of the other person involved.

The Defence of Publication in the Course of a Business

This defence concerns itself with proceedings which are brought concerning the advertising of food for sale. In that area, the Act provides a means of defence but again puts the onus of responsibility upon the defendant to prove their innocence. Specifically the Act requires:

1 that the person charged '. . . is a person whose business it is to publish or arrange for publication of an advertisement'; and
2 'that he received the advertisement in the ordinary course of business and did not know and had no reason to suspect that its publication would amount to an offence . . .'

Irrespective of whether a retailers sells 'own label' or 'branded' goods, it makes no difference to the fact that they can put forward the defence of due diligence. The requirements for the defence remain the same in both cases.

Penalties

Bearing in mind that the Food Safety Act 1990 imposes criminal liability it is not surprising to find that the penalties imposed by the courts can be in the form of a fine, a term of imprisonment or both. In the case of conviction on indictment, the Crown Court can impose an unlimited fine and/or a custodial sentence of up to two years. In addition a company or business which sells food which is unsafe and, as a result, causes injury may face the prospect of having to pay compensation. Whilst the Food Act itself makes no mention of this, there are three ways in which the requirement to pay compensation can come about.

1 The injured party could sue for civil damages under common law.
2 A consumer may be able to obtain damages under the Consumer Protection Act 1987. Although this only covers manufactured food and excludes unprocessed agricultural produce.
3 The criminal court hearing the case could make an order requiring the defendant upon conviction to pay compensation to whoever it was that suffered injury or loss as a result of the defendant's criminal act.

An additional penalty which can be a result of infringements of the Act is quite simply the financial loss which would result from having food seized and condemned. If this does happen, the owner can also be ordered by the court to meet the cost of disposal. There is one further cost to be considered in this area. If a sample from a batch of food is found to be unsafe, the courts will presume that the whole batch from which the sample came is also unsafe, unless the defendant can provide proof to the contrary. Given that any particular batch of food will be produced at one time under constant conditions this seems a reasonable premise for the courts to work from. With all of this goes the potential loss of trade due to adverse publicity. Many of the cases concerning infringements of the Food Act 1990 result in quite substantial fines. Given that such fines can be handed down by the local Magistrate it is not surprising that they are reported in the local press, often with adverse effects for the business concerned.

Time Limit for Prosecutions

Section 34 of the Food Safety Act 1990 makes a clear statement in this area, viz.:

> 'no prosecution for an offence under this Act . . . shall be begun after the expiry of:
>
> (a) three years from the commission of the offence; or
> (b) one year from its discovery by the prosecutor, which ever is the earlier.'

Given the nature of some of the possible defences afforded by the Act, once again a situation arises where some of the requirements of a well constructed quality assurance management system, in this case good record keeping, could prove its worth in contesting, if not actually preventing a conflict with the law.

Enforcement

As stated in the outline, enforcement of the Act is a matter for the local authority, either in the guise of trading standards or environmental health officers. In order to ensure that the requirements of the Food Safety Act 1990 are being complied with, the local authority officers have a number of powers, which are granted by the Act, concerning:

1 their right of entry;
2 the inspection of food;
3 issuing improvement notices;
4 the imposition of emergency prohibition orders.

Right of Entry

In accordance with the Act, local authority enforcement officers may enter premises to investigate possible offences, and in so doing they may:

(a) inspect premises;
(b) inspect processes and records;
(c) copy records and remove samples of food for subsequent analysis;
(d) compile their own visual records using both still photography or videos.

Inspection of Food

Officers may inspect food to ascertain whether or not it is safe for human consumption. In instances where the enforcement officer concerned believes the food to be unsafe the Act provides for two courses of action:

1 That the person in charge of the food is given notice, by the enforcement officer, that the food is not to be used for human consumption until such time as the notice is withdrawn and that the food in question is not to be removed unless it is to a place specified in the notice. This course of action gives the enforcement officer time to investigate the matter further.
2 If in the opinion of the enforcement officer no further investigation is needed, the food in question can be seized and the matter will then have to be dealt with by a magistrate.

In situations where a notice is given that food should not be used for human consumption, it needs to be emphasized that the issuing of such a notice automatically confers a legal requirement, in that anyone who knowingly contravenes such a notice shall be considered to be guilty of an offence. When such a notice is served by an enforcement officer, the Act requires that as soon as is reasonably practicable and in any event within a maximum of twenty-one days, the officers shall establish whether or not the food concerned complies with the food safety requirements. In instances where the food concerned does comply with the food safety requirements, the notice will be withdrawn. If however, the enforcement officer is not satisfied that the food in question complies with the Act, then it will be seized and the matter will have to be dealt with by a magistrate.

In situations where food is seized by enforcement officers, the matter then has to be put before a magistrate to be resolved. Should it subsequently be determined by the magistrate that the food which has been seized fails to comply with the food safety requirements, the Act (section 9) specifies that the magistrate must condemn the food and order that:

(a) the food be destroyed or be disposed of so as to prevent it being used for human consumption; and
(b) any expenses reasonably incurred in connection with the destruction or disposal to be defrayed by the owner of the food.

However, if a notice issued by an enforcement officer, is withdrawn, or if a magistrate refuses to condemn the food, the local authority will be obliged to compensate the owner for any depreciation in its value which has occurred as a result of the action of the enforcement officer.

Issuing Improvement Notices

In situations where an enforcement officer has reasonable grounds for believing that a food business is failing to comply with the requirements of the food hygiene or food processing regulations, the officer may serve an improvement notice upon the proprietor of the business. The issuing of this type of notice also confers legal responsibility on the recipient, in that failure to comply constitutes an offence in itself. The form that such a notice must take is laid down by the Act itself which states that an improvement notice must:

(a) state the officer's grounds for believing that the proprietor is failing to comply with the regulations;
(b) specify the matters which constitute the proprietor's failure to comply;
(c) specify the measures which, in the officer's opinion, the proprietor must take in order to secure compliance; and
(d) require the proprietor to take those measures, or measures which are equivalent to them, within such a period (not being less than 14 days) as may be specified in the notice.

In the case of (d) above, where proprietors do not take the precise steps laid out in the improvement notice, it would be advisable for them to seek the enforcement officer's opinion as to whether or not the steps being taken actually do conform to the requirements of the improvement notice.

Emergency Prohibition Notices and Orders

This type of *notice* can be served by a local authority enforcement officer, with prior reference to any court, but only in situations where a business presents an imminent risk to health. The result of the serving of such a notice is that part or all of the premises concerned will have to be closed. Once such a notice has been served the enforcement officer must, within three days, take the matter before a court, and at least one day before making such an application the proprietor of the business must be informed of the officer's intended action. In cases where the court agrees with the enforcement officer it will then grant an *emergency prohibition order* against the *premises* not the proprietor. In order to have an emergency prohibition order lifted, application has to be made initially to the enforcement authority for a certificate which states that sufficient steps have been taken to ensure the business can function without incurring a risk to the public. The enforcement authority are bound to reach a decision on such an application within a maximum of fourteen days, and providing that they concur that the risk has been removed they have a further three days in which to issue a certificate. Should the authority refuse to issue such a certificate an appeal can be made to a Magistrates Court to have the order lifted.

With regard to the actions of enforcement officers themselves, two further points need to remembered:

1 if the officers have their request for permission to enter premises denied by the occupants, management or owners, then they can apply to a magistrate for a warrant;
2 the Act requires that enforcement officers are given reasonable information and assistance. In fact failing to provide such information is an offence in itself, as is:
   (a) providing information which is known to be false or misleading:
   (b) 'recklessly providing information which is false or misleading'.

Whilst the Act appears to make every effort to prevent the obstruction of enforcement officers, it still attempts to protect the basic legal rights of the individual. To this end it makes it clear that whilst requiring that enforcement officers are given reasonable information and assistance there is nothing which actually requires a person to provide assistance or information which may be self incriminating.

Prohibition Orders

Prohibition orders can be granted by a Magistrates Court and result in a business being closed down, either completely or in part. In addition, the court can impose a ban on a particular manager or proprietor to prevent them from running a food business of either a specific kind or any kind whatsoever. However, such an order can only be imposed by a court against a proprietor that has been convicted by that court. In order to proceed with a prohibition order the court has to be satisfied that the business concerned puts the public health at risk. The nature of the prohibition order will be dependent on the nature of the risk and as a result will be concerned with one, or more, of the following:

1 where the risk is due to a particular process or treatment, the order will prohibit the use of that process or treatment;
2 in situations where the construction of the premises or the use of any equipment constitutes the risk, then the order will prohibit the use of those premises or that equipment;
3 where the state or condition of the premises constitutes the risk the notice will prohibit the use of those premises.

The requirements for lifting a prohibition order against a business are the same as for lifting an emergency prohibition order. To lift a prohibition order against an individual requires an application to the court. However, such an application cannot be made earlier than six months from the imposition of the order. In the event that an application to have a ban against an individual is refused by the court, a subsequent application cannot be made for a further three months.

Summary: the Food Act and Quality Issues

Considering what has been mentioned in this chapter concerning the Food Act, its requirements and the consequences of failing to comply, the role of quality assurance procedures and their value should be blatantly obvious. Primarily there is a fundamental role to ensure that working procedures are correctly formulated and implemented with the requirements of the Food Act in mind. Senior management, and proprietors should not be looking for a quality assurance system which will provide their organization with a cosmetic dressing. Indeed given their own vulnerability as a result of the Food Safety Act 1990, along with other areas of legislation such as the Trade Descriptions Act 1968, management should be trying to ensure that the company's procedures are rigorously implemented and maintained and, as and where necessary reviewed and updated.

General Food Hygiene Regulations 1995

Given the importance of the food industry, it is not surprising that the Food Safety Act 1990 does not stand as a single piece of legislation for those in this particular area. Whilst it is in itself a wide ranging piece of legislation which strengthens and updates the law in the areas of food safety and consumer protection, a number of other pieces of legislation concerning food remain in force. In addition, new regulations can be made under the Food Safety Act, indeed a number have come into existence since the Food Safety Act 1990 was first introduced. A great many of these regulations are product specific, covering such areas as: Fresh Meat, Wild Game and Dairy Products. Another set of regulations which were introduced in 1995 cover Temperature Control, but the one which is the most far reaching, also introduced in 1995, was the General Food Hygiene Regulations.

Outline

Briefly the General Food Hygiene Regulations came into force on 15 September 1995, introduced as a result of an EC Directive which itself aimed to create common food hygiene rules throughout the European Community.

Who do the Regulations cover?

The Regulations cover anyone who is the owner or manager of a food business. In addition they also cover *everyone* who works in a food business, with the exception of those working in primary food production which includes milking and slaughtering of animals or the harvesting of crops. Even people who clean articles and equipment which in turn come into contact with food, must comply with the Regulations.

All food businesses are covered irrespective of whether the food is sold publicly or privately, for profit or as a means of fund raising. All forms of premises are covered by the Regulations from concessionary stands and village halls to top class restaurants and supermarkets. Even vending machines are subject to the requirements of these regulations.

What constitutes a food business?

As far as these regulations are concerned, every process which deals with the sale or preparation of food can be classed as a food business. Such activities include: manufacturing, processing, preparation, handling, packaging, storage, transportation, distribution, supplying or selling.

The penalties involved

As with the Food Safety Act itself, the Regulations are enforced by local authority environmental officers. Potential penalties for offences committed in contravention of the Regulations, include the liability to be fined, serve a term of imprisonment or both.

The Main Requirements

Schedule 1 of the Regulations is separated into ten different chapters which cover:

1 general requirements for food premises;
2 specific requirements in rooms where food stuffs are prepared, treated or processed;
3 requirements for moveable and/or temporary premises;
4 transportation;
5 equipment requirements;
6 food waste;
7 water supply;
8 personal hygiene;
9 provisions applicable to food stuffs; and
10 training.

Whilst these are industry specific, many of the requirements could be incorporated within the bounds of a quality assurance management system, conforming to such as BS EN ISO 9000, but specifically written for a food business.

Very little of what is contained within the ten chapters is new to the food industry. What could well be new is how they could be encompassed within the procedures of a quality assurance management system rather than being allowed to stand alone as procedures created specifically to satisfy the needs of the General Food Hygiene Regulations 1995.

Comparison of the requirements of BS EN ISO 9000 with the requirements of the General Food Hygiene Regulations produces some very obvious parallels. For example, within the Quality Standard there is a requirement to control processes which themselves can have a direct effect on quality. Part of this requirement concerns the suitability of the working environment. Contrast this with the Food Hygiene Regulations which in the first chapter of Schedule 1 makes reference to the layout, design construction and size of premises to ensure that they permit good hygiene practices. Again within the same section of the Standard there is reference to the use of suitable equipment, whilst Schedule 1 chapter 5 of the regulations specifically concerns equipment which comes into contact with food.

The comparisons made above concerned only one small part of BS EN ISO 9000 whilst they have looked in part or in whole at two chapters of Schedule 1 of the Regulations. Further comparisons, whilst beyond the scope of this book, are very easy to make, indeed they could quite easily form the basis for a publication of their own.

One aspect of the Regulations which is new, and does bear comparison with the Quality Standard is that of training. There is within the standard a specific requirement to ensure that training is undertaken by all those whose activities have an effect or bearing on the quality of the product or service. Chapter 10 of Schedule 1 of the Regulations concerns itself specifically with the need to provide training in food hygiene. However, the Regulations themselves do not specify exactly the form that the training and/or supervision should take. Rather it leaves it to each individual business to identify its own training requirements, which is very much the case with regard to the conditions specified by the Quality Standard concerning training.

The identification and control of food hazards

This is the new requirement within the regulations. What it requires is that businesses identify potential hazards – those things which may be harmful – and introduce controls in such areas which may prove to be critical to food safety. In addition:

- it is a requirement that controls are regularly monitored to ensure that they are operating effectively;
- controls need to be reviewed and maintained; and
- periodically, and at such times as food operations change, the assessment, control and monitoring procedures need to be reviewed. In this area the requirements of the Regulations work very much hand in hand with the operational requirements of a quality assurance management system.

With regard to the regulations an example of how the regulations could be conformed to in the area of food preparation, might be:
  (a) identifying potential hazards, e.g. bacterial growth or contamination from chemicals or pests;
  (b) the implementation of controls such as limiting handling times, ensuring that equipment used is always clean and suitable for the task,

insisting on good standards of personal hygiene and ensuring that the premises are kept in a hygienic condition;

(c) monitoring might take the form of visual inspection and the implementation of cleaning schedules.

Records

Surprisingly the Regulations do not require the keeping of written records of analysis and monitoring procedures even though the existence of such records could well help in demonstrating compliance with the Regulations. However, such procedures could well be encompassed as a normal part of quality assurance procedures, which if kept and maintained properly could automatically demonstrate compliance with the Regulations.

Hazard analysis and critical control points

What some businesses may find confusing and frustrating is that the Regulations do not themselves specify a particular system that must be used in identifying and controlling potential hazards to food safety. A formal system known as HACCP (Hazard Analysis and Critical Control Points) does exist and in some respects is similar to the Failure Mode and Critical Effects Analysis (FMCEA) which is used by quality assurance personnel in manufacturing industries. Individual businesses can seek advice from their local authority environmental health officers. However, it needs to be borne in mind that the degree of complexity of such a system will very much depend upon the nature and complexity of the individual businesses concerned. It also needs to be borne in mind that irrespective of the system used, the ultimate responsibility for complying with the requirements of the Regulations remains with the individual business concerned.

Summation

If food companies and businesses are going to involve themselves in a quality assurance management system, it should be for the same reason for them as it should be for any other company in any other industry, namely to provide a working tool aimed at ensuring that their practices consistently conform to a recognized and independently verifiable international standard. It would seem to makes sense that in so doing, the quality assurance procedures should where possible be mindful of, and incorporate, the requirements of relevant legislation. The difficulties in the creation of a quality assurance management system in accordance with such as BS EN ISO 9000 should never be dismissed. In addition, giving due regard to the requirements of the law within that system would seem to compound the problem, and no doubt in some instances it does. However, the burden must be considerably eased for those in the food industry by legislation which has clear objectives as to what is required as a matter of law. Given the brief comparisons which have been drawn above, together with the many that can be drawn concerning the law and Quality Standard BS EN ISO 9000, it is in the opinion of the authors made relatively easier, in some parts, for the food industry due to the requirements of the Food Safety Act 1990 and the General Food Hygiene Regulations 1995.

8

BS EN ISO 9000 Quality Systems

Issued in 1994 and formerly known as BS 5750, the BS EN ISO 9000 series of quality systems have been referred to many times in the previous chapters. Those new to the quality world could be forgiven for assuming that they are the only form of quality standard in existence, which is most certainly not the case. Companies concerned with, for example, the production of military defence and aerospace equipment need to concern themselves with quality standards which have been generated to suit the specific needs of their own particular industries. Similarly, component companies and suppliers to such as the Ford Motor Company, will need to conform to that company's own particular requirements, which in the case of Ford is their Q1 System. However, that does not necessarily mean that such organizations do not have a role or need for BS EN ISO 9000. What this series of standards sets out to do is to provide quality systems which can be adopted by a wide variety of business serving any number of different industries. Companies which produce their own quality assurance management systems, which in turn conform to the requirements of the British Standard, can upon independent assessment, become registered as *Firms of Assessed Capability*. The number of companies which undertake this route increases every year, as does the variety of different business activities from which they come. Originally branded as being biased in favour of manufacturing industries, the Standard has now been adopted by a number of diverse businesses which include doctors' surgeries, solicitors' practices and even a public house. It is because of the adaptability of the Standard, to such diversity of applications, that it has been referred to extensively throughout the preceding chapters.

The Standard in itself has been the source of a whole array of publications, with additional books being published concerning its application to a variety of business situations. It is not therefore the aim of this chapter to try to recover that ground, nor is it the aim to try to show how it can be applied to every business within the industries previously mentioned. To do so would be an enormous task in itself, with large parts being industry specific and as a result irrelevant to a great many readers. The aim of this chapter is to try to show how the various parts of BS EN ISO 9000 can have specific relevance to the aspects of law which have been covered in the preceding chapters, and as a result show how particular aspects of legislation might be borne in mind and catered for, when quality assurance management procedures are compiled or reviewed. That being the case the following is listed under the section numbers and headings to be found in BS EN ISO 9001, the specific requirements for each section being found within the Standard itself.

S.4 Quality System Requirements

As with all the sections of the Standard, the section entitled 'quality system requirements' contains a number of different subsections, each of which could be considered in such a fashion that they are given importance with regard, not only to quality, but also to the requirements of legislation pertaining to a particular type of business or industry. For example, within subsection 4.1.2.2, the Standard specifies a requirement for the identification and provision of adequate resources, a requirement within which it includes trained personnel. For those in the food and catering industries, these requirements can be related automatically to the requirements of the General Food Hygiene Regulations 1995 – see Chapter 7. In brief these regulations themselves make it a requirement that all food handlers must receive adequate supervision, instruction and or training in food hygiene. A similar situation occurs with the *Health and Safety at Work etc. Act 1974* which makes it a duty of every employer to provide instruction and training so as to ensure the health and safety of employees.

These brief examples alone give a good indication of how the requirements of the Quality Standard can be linked to the requirements of legislation. What follows is an attempt to indicate how some of the other requirements of the Standard could be related to the legislation covered in the preceding chapters.

4.1.1 Quality Policy

Also part of the quality system requirements (section 4), as the heading suggests, is a requirement for a company to define its policy with regard to quality. On the face of it the Standard itself appears to have little apparent connection in this area with legislative requirements. However, the Standard does stipulate that the quality policy should have relevance to the needs and expectations of customers. Given that customers have the legal right to expect safe products or services, surely the formulation of quality policy itself provides an ideal opportunity for a company to make clear its intention to provide such products or services, whilst at the same time taking the opportunity to review its legal corporate obligations and make management and employees aware of their own individual and collective responsibilities.

4.1.2 Organization

Within this section there is created the specified need to establish clearly areas of responsibility and associated authority. This should always be done by job title or description, as opposed to identifying individual personnel by name. Obviously the role of quality manager will be cardinal, although the role itself will almost certainly vary from company to company. Larger organizations for example may well have an individual dedicated solely to that function, whereas smaller businesses may combine the role of quality manager with that of another specialism, the usual combination being that of quality and training or quality and safety. Given that only the larger organizations will be able to afford a qualified solicitor or legal executive as part of the staff, with smaller companies seeking legal advice from specialist legal firms, there is nevertheless a need in all companies for someone with an understanding of the legislation relevant to the company. So this becomes another possible, and in the opinion of the authors, desirable combination – quality and the law. Indeed it is the very reason for the existence of this book.

4.1.3 Management Review

This is an obligation specified by the Standard to ensure a review of the quality system is undertaken by senior management to ensure the continuing validity and effectiveness of the company's quality management system. Whilst the Standard itself does not specify the time interval at which such reviews should be carried out, it does make it a requisite that the interval does need to be defined. Current general practice would seem to make this interval not longer than once every twelve months. Such a review should be examining a number of different areas, such as:

- the findings and effectiveness of quality audits;
- customer feedback;
- changes in technology; and by no means least
- any changes in legislation which itself will have a bearing on the business, its products or services.

S.4.2 Quality Systems

This section focuses the need to ensure that the product or service conforms to specification by requiring that a documented quality system is created and implemented as a means of meeting this need. An oft heard criticism of the Standard is that it does not specify exactly what is required, which when compared to other standards such as BS 308 'Engineering Drawing Practice' is quite true. After all, the latter goes to great lengths to specify exactly what is required with regard to everything from the sizes of drawing paper to the style and thickness of line types which should be used for different aspects on drawings. So compared to this type of standard the criticism of BS EN ISO 9000 is somewhat justified. What needs to be borne in mind when making such a comparison is that engineering drawing is not only a national but also an international method of communication and as such needs finite definition. By comparison, whilst the BS EN ISO 9000 series is also an international standard, the method by which individual quality assurance management systems conform to the Standard will be as different as the companies trying to implement such systems. Indeed the Standard recognizes this diversity within section 4.2.2 (quality system procedures), and makes it clear that quality procedures which are drawn up with the aim of conforming to the Standard will themselves depend upon the specific requirements of the individual business concerned. It can be argued that companies in the same line of business could or should have a degree of commonality, and this may well be the case. This notwithstanding, unless two or more companies producing identical products or services belong to the same group, with the same management structure, its highly unlikely that their needs will be identical. At some point in the equation, be it with regard to the nature of the work and methods used, or the skills and training required, similar companies with apparently identical products will have differing needs. Hence the Standard sets out specific requirements for all, whilst at the same time acknowledging that these requirements are likely to be met in different ways.

This lack of apparent absolutes within the Standard has undoubtedly been the cause of much frustration amongst those charged with the task of creating and implementing quality producers. Indeed many smaller companies have turned to a consultant who has offered an 'off the peg' quality system tailored to suit individual businesses. Undoubtedly in some cases this has proved to be an advantageous approach, but as in all things what needs to be borne in mind is the legal maxim *caveat emptor* – buyer beware. The benefit offered by apparent the lack of absolutes in the Standard is that the requirements of the law can be built into the quality procedures and thus adopted as a matter of everyday practices, thereby killing two birds with one stone. Whilst this was not necessarily in the forefront of the minds of those that created the Standard, as these pages have attempted to show, it could have been worthwhile on numerous occasions.

S.4.3 Contract Review

The process of reviewing contracts or orders should be a fundamental part of any business, irrespective of its adoption of quality assurance management procedures. The need to establish that contracts which are entered into are fully understood, unambiguous, capable of being fulfilled, and in keeping with the customer's requirements are paramount to any company's success. What BS EN ISO 9000 sets out to do is ensure that formal procedures exist to ensure the process of reviewing contracts is carried out prior to the acceptance of any order, or preferably prior to the submission of a tender. Failure to instigate and carryout such procedures will of course constitute a non compliance when compared with the Standard, although the absence of, and the failure to implement such procedures do not in themselves constitute a breach of English law. However, as can be seen in Chapter 2, the case of *Butler Machine Tool Company* v. *Excello Corporation* (1979) provides an example of how the failure to undertake contract review procedures thoroughly can prove to be very expensive, when that failure leads to litigation.

S.4.4 Design Control

Comprising of nine subsections, design control sets out to ensure that all aspects of the design process are covered. To this end there is one direct reference, and one indirect reference to a company's legal obligations within the various subsections. Subsection 4.4.4 – Design Input – specifies the need to ensure that statutory and regulatory requirements are identified and as a result obviously observed. Similarly subsection 4.4.5 – Design Output – makes it a requirement to identify those aspects of the design which will have an effect on the safety of the product. Further, within subsection 4.4.6 – Design Review – there is a specific requirement for such reviews to be carried out by personnel who represent all of the various functions which are included in the design stage. An obvious need then is for someone within the organization to be versed in the areas of legislation which appertain to the company's products or services. Readers should be mindful of this last statement. Design control is not solely restricted to industrial manufacturing industries. A design could, for example, be the creation of a new recipe. Services industries of any type, if registered as a firm of assessed capability to BS EN ISO 9001, have to incorporate design control, and in so doing will need to be mindful of the legal requirements which appertain to their own specific industry, as well as those such as the Health and Safety at Work etc. Act which applies to all.

S.4.5 Documentation and Data Control

Documentation is one of the key issues within the Standard, virtually all sections requiring that documented procedures are created and maintained. No wonder then that the Standard makes a specific requirement as to how documentation and data is controlled with regard to its original approval and issue, and also with regard to the manner in which changes to such documentation are made. Specific requirement is made concerning the adequacy of documentation and the need for such adequacy to be ensured by authorized personnel. Given that this book advocates the need to combine legal necessity with the requirements of BS EN ISO 9000 it seems obvious that amongst those personnel who are authorized with documentation control is at least one person who is sufficiently aware of the law that they can try to ensure that the legal necessities are not overlooked.

S.4.6 Purchasing

This section of the Standard is subdivided into three parts and concerns itself with:

(i)   the evaluation of subcontractors (4.6.2);
(ii) purchasing data (4.6.3); and
(iii) the verification of purchased product (4.6.4); which is itself subdivided into:
    (a) supplier verification at subcontractor's premises (4.6.4.1); and
    (b) customer verification of subcontracted product (4.6.4.2).

The purpose of the section is in essence to ensure that the right type of product is ordered, to the right specification from a subcontractor who has the capability to meet the requirements of the product ordered. The provision to undertake verification of the subcontractor's premises and product by the supplier and customer respectively, provides the opportunity to ensure that the right product is indeed forthcoming. Nothing within this section is anything other than sound business practice aimed at ensuring that subcontracted items are right first time. The section also provides the opportunity to address at least one legal concern.

Consider the concept of joint and several liability which exists as part of the Consumer Protection Act 1987. One company – the supplier – may well be assembling finished products which are made up of components produced by a second company – the subcontractor. Take a hypothetical situation where faulty components are assembled into a product which subsequently causes injury to the eventual purchaser. Who is liable? According to the law, in this hypothetical case, the company assembling the product and the subcontractor are both jointly and separately liable, seemingly unfair to the company assembling the product as, in this example, the blame would, to the lay person, seem to rest with the subcontractor. Seemingly even more unfair is the fact that the eventual vendor, putting their own brand label on the product, would in this case also be held responsible.

The Consumer Protection Act is, it should be remembered, about protecting the consumer. The means for the assembler and vendor to seek redress from the subcontractor is provided by means of civil law, and it could be that in seeking such redress that section 4.6 (Purchasing), if well worked, could prove to be a valuable legal aid. In essence the supplier will try to establish that they have themselves suffered as a result of the actions of the subcontractor, in that the subcontractor did not fulfil their own contractual obligations for whatever reason. In so doing the supplier will be trying to demonstrate that they themselves were not negligent in their dealings or that they exercised due care or due diligence, dependent upon the circumstances and the products involved. To aid in this, the quality procedures for purchasing (4.6) if properly constructed and implemented will show that:

(a) the subcontractors were properly evaluated with their regard to their ability to fulfil the requirements of the contract;
(b) that the correct controls over the subcontractor were in place and duly exercised; and
(c) the purchasing data supplied to the subcontractor was complete, accurate and specific with regard to what was actually required.

The supplier will need however, to be careful with regard to verification which may / should have taken place at the subcontractor's premises. These things done, which they should as part of normal practice, especially if that practice is part of a quality system which conforms to BS EN ISO 9000, could well provide the valuable legal assistance mentioned above.

S.4.7 Control of Customer Supplied Product

The Standard makes it quite clear that in situations were items (product) are supplied by the customer, to the supplier, for incorporation into subsequent products, the onus of responsibility lies with the customer to ensure that what they supply is acceptable. However, the Standard requires that such acceptability is verified and subsequently maintained, and that any unsuitability, for what ever reason, is reported back to the customer. Justice would hardly be served if the customer subsequently brought a successful legal action, under the laws of contract or negligence, because they received faulty products, which were a result of items which they themselves supplied, but were never identified as such.

S.4.8 Product Identification and Traceability

Firstly, looked at from the standpoint of an issue concerning quality, it ought to be obvious that without correct product identification, tracing the source or origin of a particular item, be it an ingredient, component, subassembly or complete product, becomes extremely difficult. This will be so whether the item is produced internally, by either one of a number of machines and/or operators or externally by dual source suppliers. If we are following the ethos that problems are to be learned from and rectified at source, little can be achieved if this source cannot be identified. This surely is a situation which faces all organizations and as a result it seems only common sense that procedures concerning product identification and traceability should be in place irrespective of whether or not a company embraces the requirements of BS EN ISO 9000.

Looked at from the legal standpoint, the ramifications are equally obvious. Although it can be seen that the areas of product identification and traceability are most definitely interlinked, there is at the same time, in legal terms, the possibility of these two areas being giving rise to two separate issues. Firstly, incorrectly identifying a product which subsequently reaches the customer could well give rise to action under the Trade Descriptions Act, with the prospects of the penalties which that can incur – see Chapter 6. Secondly, consider the requirements of the Consumer Protection Act 1984, which stipulates that liability is joint and several. Being able to trace a faulty item to a subcontractor may not prevent action being brought under the Consumer Protection Act, but it may well provide the means of pursuing redress against the supplier, by means of contract law or the tort of negligence. So by complying with the requirements of what is a very short section of the British Standard could at some point in the future negate some, if not all, of the financial cost of a legal action, if not actually providing a means of preventing that legal action from arising in the first instance.

S.4.9 Process Control

With the introduction of BS5750 – the forerunner to what is now BS EN ISO 9000 – the general misconception was the Standard was aimed solely at manufacturing organizations, even though the Standard itself pointed out that the word product should also be read as service. However, being as it was, and to some degree still is, misconstrued in this fashion, the section on process control was also largely seen as the area which had the most direct effect upon the quality of a manufactured product. Nevertheless, the Standard is quite specific in that process control is aimed not only at production but also at installation and servicing. That said, there is obviously considerable scope for poor practices to creep into these areas, practices which in turn will have a detrimental effect upon the quality of a product or service. What may be not quite so obvious is that within the area of process control, there is also considerable scope for a company to come into conflict with their statutory duties. Consider poor product quality that could create unsafe situations for the eventual purchaser and user, and thereby create a situation where a company is failing in its obligations under the Consumer Protection Act 1984, or the General Product Safety Regulations 1990. Take a similar situation in the food industry where a poorly conceived or executed procedure could be seen as failure to exercise due diligence, and as a result bring about legal action under the Food Act 1990. Similarly poor procedures could be seen as a failure to exercise due diligence, and as a result negligence. Worse still would be the total absence of process control procedures, or working instructions, which was in essence the reason for the successful prosecution in the case of *Denard* v. *Smith* (1990) (see Chapter 6 – Fair Trading). There is in reality the potential, due to poor process control, to come into conflict with all the legislation mention in this book, and probably a great deal more besides. One further instance worthy of mention would be a situation where process control procedures were so badly conceived or implemented that they put the company's own workforce into danger, and as a result contradicted the requirements of the Health and Safety at Work etc. Act 1974.

It would be quite wrong to suggest that one section of the Standard has a greater relevance than any other, for they are designed to work in unison as part of a comprehensive system. It would also be extremely inefficient business practice to work a so-called quality system which was centred round inspecting out non conformity, although in the past this was quite often the case. The particular importance of this section of the Standard is that it recognizes that in a less than perfect world, quality is not always going to be *right first time*, non conforming product (substandard service) will occur. As a result procedures need to be in place to detect non conformity – *inspection and testing* – and once detected additional procedures need to exist which will control the non conformity and prevent it, in whatever guise, from reaching the customer. This is nothing less than sound business practice. Preventing non conforming product from reaching the customer prevents the possible cost, and as a result loss of profit, which arise from warranty claims, recall, replacement all of which are quantifiable, together with the potentially more costly loss of reputation. In addition it has the benefit of preventing non conformity bringing the supplier or manufacturer into conflict with the law, which may result in prosecution, and or the payment of damages, but will in either event lead to a tarnished reputation and regardless of outcome incur costs of an internal or external nature.

S.4.10 Inspection and Testing

Ensuring that the product specifications are complied with is a prime requirement of this section of the Standard, as is ensuring that verification of conformance is carried out in accordance with the documented procedures or a quality plan. It should be noted that here are not one, but two areas of a quality system which could cause conflict with statutory requirements.

Consider again the case of *Rotherham Metropolitan Borough Council* v. *Raysun (UK)* (see Chapter 6 – Fair Trading). Suggestion could be made that Raysun (UK) did in part conform to one of the requirements of this section of the Standard – namely 4.10.2 Receiving Inspection and Testing. The company placed a requirement with their Hong Kong supplier that they should be notified of any adverse test results which indicated toxicity in the wax crayons which they were importing. Having not received any adverse reports, the company conducted their own analysis to ensure that the crayons were indeed as they were advertised – poisonless. In so doing they were in essence fulfilling the requirement of this part of the Standard. The downfall of the company was the amount of inspection they carried, which according to the court '... did not indicate the standard of care required by the statutory provisions'. Note that it was not the failure to carry out what the Standard requires in terms of verifying that a product adhered to its specification, but a case where either:

(a) the inspection was not carried out in accordance with a quality plan; or

(b) that the quality plan itself did not specify a level of inspection that could itself 'indicate the standard of care required by the statutory provisions.'

There is the possibility of a third situation, namely that where no quality procedures or quality plan existed. However, it is most unlikely that this situation would go unnoticed in any organization being audited to the standard required to become a *firm of assessed capability.*

S.4.11 Control of Inspection, Measuring and Test Equipment

The control of this type of equipment forms an obvious area of concern in manufacturing industries when considering quality. Ensuring components are made to the specifications in order that they can be correctly assembled or replaced is fundamental to manufacturing companies. Failing to produce components to the correct specifications can lead to the problems of replacement, warranty claims with their associated cost such as additional manufacturing and materials, for the replacement parts, and the manpower involved with the manufacturing, re-inspection, complaints investigation and delivery etc. These things are common enough, being a familiar problem to most organizations. But consider also the potential legal implications of supplying components which actually fail to meet their specification due to what is only some inaccuracy in the inspection, measuring or test equipment. Implications such as:

1 Civil actions brought for breach of contract, components not to specification upon delivery could well have a snowballing effect causing further delays in the delivery of the items to which they were to be installed. Alternatively the out of specification components could be part of a customer's own production or manufacturing process which is subsequently delayed, causing loss of production, subsequent loss of revenue and the demand for compensation.

2 The out of specification components or products causing injury, which could in turn result in claims of negligence – see Chapter 3 – which may lead to the award of damages in a civil court.

3 A claim for damages due to the product being determined as unsafe with regard to the Consumer Protection Act 1984 – see Chapter 4

4 Prosecution as a result of the Consumer Protection Act 1984 or the Unsafe Product Safety Regulations 1990 with all the penalties and cost which that itself may entail.

The presence of well constructed and correctly implemented quality assurance procedures as required by BS EN ISO 9000 could well go a long way to ensuring that the potential legal implications arising from the failure to control inspection, measuring and test equipment do not arise.

Look for a moment at an area other than manufacturing and the necessity for having good control of inspection, measuring and test equipment is even more obvious. The Weights and Measures Act 1985 – see Chapter 6 – concerns itself specifically with quantitative issues. Indeed the Act specifically requires that all weighing and measuring equipment used for trade is maintained within specified limits. Failure to do so could well result in criminal prosecution. This is a further instance of how meeting the requirements of the Standard can go a long way to ensuring the requirements of the law are also met.

S.4.12 Inspection and Test Status

This section of the Standard creates a requirement to indicate the inspection and test status of a product, which itself has a direct comparison in law with the requirements of the Trade Description Act 1968 and the Weights and Measures Act 1985 – see Chapter 6. Obviously from the point of view of both quality assurance and the law, the primary requirement is that any indicator of inspection or test status should accurately and truthfully reflect that status. This is a requirement which in itself necessitates that the quality procedures which are created concerning this area are well constructed and correctly implemented. This may in turn provide the additional benefit of assisting a defence against a possible claim of negligence – see Chapter 3 – by indicating that the producers had exercised due diligence and as a result not breached their duty of care.

S.4.13 Control of Non Conforming Product

As part of a quality management system, this section of the Standard concerns itself with preventing that which is sub standard, whether it be machine parts or food ingredients, from being used or installed. As a result, it is of automatic benefit in trying to ensure that a company's products or service does not breach statutory requirements. The control of non conforming product must by both inference and necessity mean that the producer is attempting to exercise due diligence in trying to ensure that such products or forms of service do not reach the customer. If well established and equally well enacted, procedures of this nature could well prevent dangerous or unsafe items from reaching the point of sale, hazarding the safety of the consumer, and resulting in costly litigation. The benefit to be gained from implementing this section of the Standard was no doubt originally conceived to be that of ensuring customer satisfaction, or at least preventing dissatisfaction. What it also does, is to serve as an ideal example of how, by using a quality management system which itself conforms to the British Standard, the requirements of the Standard can have the additional benefit of helping to ensure that a company does not breach its legal obligations.

S.4.14 Corrective and Preventive Action

Prevention is better than cure according to the proverb so perhaps this section would have been better entitled Prevention and corrective action. In reality this section ought to be the essence of the approach to quality, the prevention being the right first time objective, whilst the corrective action is an acknowledgement that things do not always go according to plan, but nevertheless provide opportunities to improve for the future.

The irony is that reputations can and are advanced more by the speed and effectiveness of corrective action than they are by preventive actions which result in the product or service being right first time. This statement is particular true at the point of sale when dealing with individual customers. It is a well known characteristic of the English that they do not complain well. Take for example a restaurant that serves up a dish which is not to the customer's satisfaction. Either the customer will quietly accept what they have been given, preferring not to make a fuss, or alternatively immediately become adversarial when making their complaint. In either event the restaurateur is presented with a potential no-win situation. Whilst the no fuss scenario seems to be wholly in favour of the restaurant's management, it presents them with no opportunity to rectify the situation as it occurs or to learn from the experience thus preventing it from being repeated. It has the additional disadvantage that once away from the restaurant the customers will undoubted recount their poor experience, thus tarnishing the establishment's reputation. Such instances undoubtedly have an effect upon a business, upon its potential takings and as a result its profits. Unfortunately, although theories abound as to how much damage is done to reputations by such incidents, whilst they remain unquantifiable a great many businesses will simply cast them aside. The second scenario, that of adversarial complaint, seems to present the restaurateur with the worst option. Poorly handled, a public complaint will effect the other customer who in turn will recount the incident, which has a snowballing effect with each retelling harming the restaurant's reputation further.

Perversely it is the second scenario which affords the restaurant's management with the best opportunity. Whilst it should obviously be accepted that providing an unsatisfactory dish is unacceptable, an immediate complaint provides an immediate and first hand source of information concerning the quality of the product, information which should in turn be put to good used to prevent any further reoccurrence. In addition a public complaint handled properly, by individuals who have the ability and have had the necessary training to enable them to do so, can produce a veritable public relations coup, with the customer recounting how well their complaint was dealt with. In such cases customers tend to forget that the situation should not have arisen in the first instance.

As regards conforming with the law, regardless of the industry or business concerned, the emphasis has obviously got to be placed upon prevention. The possibility of an out of court settlement should not be misconstrued as the no fuss scenario. It still incurs numerous costs, reduces profits and damages reputations. Furthermore, where criminal liability is concerned out of court settlements are not an option.

S.4.15 Handling, Storage, Packaging, Preservation and Delivery

By its very title, this section of the Standard suggests immediate connotations with both the Food Act 1990 and the General Food Hygiene Regulations 1995. The subsections of the standard entitled 'Handling' – 4.15.2 – and 'Storage' – 4.15.3 – both concern themselves with the need to prevent deterioration of the product, an obvious requirement of any organization dealing with food. Similarly the subsection 4.15.5 'Preservation' also specifies what is a prerequisite for those in the food industry. Whilst subsection 4.15.4 'Packaging' addresses the need not only to ensure that packaging is as it should be but also that markings are correct, the later also being a requirement of section 15 of the Food Act.

Moving away from the food industry, the requirements of subsection 4.15.4 with regard to correct labelling also have connotations with legislation such as the Trade Descriptions Act 1968 (see Chapter 6) which makes it an offence to apply a false description.

The final subsection – 4.15.6 – concerns itself with 'Delivery.' Again obvious associations can be drawn with food and the need to ensure it does not become contaminated in any way. In terms of legal necessity the implementation of a QAMS which includes this section could be beneficial to a number of industries, not just those concerned with food. Consider for a moment that the method of delivery along with methods of handling, storage and packaging may be damaging to the product in such a way as to not only reduce its value, but also be so damaging as to render the product injurious to health. That in turn could have legal consequences as a result of the Consumer Protection Act 1987 or the Product Safety Regulations of 1990. Further, the method of delivery along with the manner of packaging could be part of the contractual agreement between the seller and customer, which in turn could have repercussions under the civil law regarding contract (Chapter 2). So whilst on the face of it this subsection may seem to concern itself solely with matters appertaining to quality, falling short in the areas covered could well have serious legal ramifications of both civil and criminal nature.

S.4.16 Control of Quality Records

The Standard has three basic aims with regard to the control of quality records:

1 the creation and implementation of procedures which are themselves able to demonstrate that specified requirements are adhered to and that the quality system is operated effectively;
2 procedures are implemented to ensure that quality records themselves are identified, assembled, indexed, stored and made accessible; and
3 quality records are legible and maintained in a in a manner which ensures that they will not be subject to damage, lost or otherwise rendered unserviceable.

Within this section is the stipulation that such records need to be made available in situations where their availability is a matter of contractual agreement. Whereas previous sections of the standard have concerned themselves with the need for procedure for specific tasks, such contract review, design control, purchasing data etc., this section concerns itself with the efficacy of procedures which themselves ensure that those records are maintained and retained within a system which ensures that they remain of worth. There is after all little point in having procedures which themselves help to prove that a business meets its legal obligations with regard to: its 'duty of care' (see negligence – Chapter 3) or 'due diligence' (see Food Act

– Chapter 7) if records of these procedures could not be produced as evidence. Similarly it would be asinine to claim the defence of warranty in a case brought under the Weights and Measures Act (see Chapter 6), if one of the linchpins of the defence – the certificate of conformance – could not be produced because it had been stored in a manner which had allowed it to become lost, or damaged to the point where it became illegible.

Quality systems are often criticized for becoming paper mountains, with ensuing problems of control and storage etc., this criticism being especially prevalent in organizations where QAMS are poorly implemented. Considering that the Standard itself makes note of the fact that such records can be stored on any manner of media, there seems to be little reason why the majority of companies should not take advantage of the technology revolution to control their documentation. The ramifications for failing to do so are twofold:

1 a non-conformance with regard to the standard itself; and
2 the opportunity to assist in:
   (a) meeting legal obligations; and
   (b) providing potential evidence in the mounting of a legal defence.

S.4.17 Internal Quality Audits

The necessity for internal quality audits is twofold. Firstly they are conducted to ensure that the quality system is being applied correctly, and secondly that reported or suspected defects in the system are eliminated. Whilst the former is of obvious importance, it needs to be remembered that a quality system is not only a management tool but also a living entity in its own right, and as such it could well be born with flaws which need to be eradicated. Similarly it should be allowed to develop as circumstances dictate. Better that rules should be made for the guidance of wise men rather than for the obedience of fools.

As a prerequisite of the Standard, internal quality audits themselves are not aimed at ensuring that a company does not find itself at odds with the law. However, if, as has constantly been advocated in the pages of this book, procedures are created which are mindful of a company's legal obligations within the areas to which those procedures apply, it then follows suit that if those procedures and their implementation are themselves the subject of an internal audit, the company operating those procedures will be afforded the opportunity of ensuring that they are working within the requirements of the law as well as within the requirements of BS EN ISO 9000.

S.4.18 Training

With regard to BS EN ISO 9000 the necessity for trained personnel has already been mentioned above in section 4 – Quality System Requirements, with specific reference made to the requirements of the *General Food Hygiene Regulations 1995*. This particular piece of legislation, although being industry specific, aligns very closely with the Standard itself, the need for each company to establish its own training requirements, and to provide for the necessary training of personnel, being common ground to both BS EN ISO 9000 and the General Food Hygiene Regulations. There is also an obvious parallel to be drawn between the Standard's requirements for the training of personnel who can affect quality, and the law's requirements for the training of personnel who can have an effect on food hygiene.

Another situation which arose partly because of the lack of training was the case of *Baxters* v. *Manley* (1995) – see Chapter 6 Fair Trading – a case which was brought under the Trade Description Act 1968 and the Weights and Measures Act 1963. Part of the reason their defence failed was because the court held that there had been insufficient training.

Numerous other instances where the law has been breached can be directly linked to the lack of training, or even, in some cases to the failure to even identify fundamental training requirements, which is a basic requirement of BS EN ISO 9000. The two examples above are barely the tip of the iceberg. Irrespective of the nature of the business, all industries or businesses will also be able to make their own comparisons between the requirements of the Standard and the law, the Health and Safety at Work etc. Act 1974 being an obvious point of concern for all. But training or the lack of it is a subject that goes much further than that.

S.4.19 Servicing

The shortest of all the sections in the Standard, this section will only be a prerequisite in quality systems where servicing is a specified requirement. The section stipulates the need to ensure that where servicing is carried out it is performed in accordance with the requirements specified. The legal implications here are twofold. Failing to perform the servicing correctly may give rise to accusations of negligence or alternatively of a failure to fulfil contractual obligations as and where they exist. Secondly, procedures and work instructions concerning servicing requirements may themselves come under scrutiny especially in cases where their existence is put forward as part of a defence where a breach of duty of care, or failure to exercise due diligence is suggested.

As a result, the section although small in itself, lays down the same fundamental requirements as all of the others, in that where servicing is carried out, quality procedures need to be both well constructed and equally well executed with both a view to the quality of the product and the requirements stipulated in law.

S.4.20 Statistical Techniques

The Standard is brief but very specific in this area. Firstly it establishes the need for identifying areas where statistical techniques should be applied. Secondly it requires that documented procedures are created and maintained to ensure such techniques are implemented. What the Standard does not state is what form such statistics should take, nor does is specify the degree to which they are applied. Obviously to do so, given the diversity of application of the Standard, is well beyond its own scope. At this point it might well be worth remembering the remark made by Disraeli – 'there are three kinds of lies: lies, damned lies and statistics.' Overly sceptical, maybe? But add to this the fact that statistical techniques are not always chosen or applied correctly and you have the possibility of a case like that of *Rotherham Metropolitan Borough Council* v. *Raysun (UK)* (1988) – see Chapter 6 Fair Trading. This was a case essentially concerning public safety. The case brought against Raysun (UK) concerned alleged contravention of the Trade Descriptions Act 1968, the Consumer Safety (Amendment) Act 1986 and the Pencils and Graphite Instruments (Safety) Regulations 1974. Specifically the case concerned the toxicity of wax crayons. Agents in Hong Kong were required to report on any adverse analysis of the crayons, but failed to do so. The sampling plan used was one packet from 10,000 dozen. It was ruled that in this country such a sample would not, in the absence of evidence showing the likelihood of the whole batch being similar, amount to the exercise of reasonable precautions or due diligence by the English sellers of toxic crayons, an example for all of the necessity not only to identify the need for statistical techniques, but also to ensure that they are themselves suitable and correctly employed.

Conclusion

The implementation of quality assurance management systems into companies has over the years been greeted in a variety ways which have depended upon the attitudes of those groups and individuals which have come under its influence. There have been, and undoubtedly will continue to be the cynics who see the Standard as something to which 'lip service has to be paid'; a flag waving exercise which results in little more than a certificate to hang above the company mantelpiece. Others will see it as an evil necessity thrust upon themselves in order to meet the demands of larger organizations and thereby retain their confidence and custom. Many more, and hopefully the vast majority, will see the Standard for what it sets out to be, a series of common sense measures, admittedly some of which are in need of interpretation to suit individual circumstances, set out in a logical and organized fashion. The creator of paper mountains is one of the most commonly heard criticisms, and indeed the need for written procedures goes some way to giving credence to the claim, but is that the case? One of the most commonly seen faults in companies trying to implement such a system, is that they do not follow the simple maxim of: 'say what you do and do what you say' which in turn creates situations of non compliance with either the standard itself, or a company's quality system created in accordance with the requirements of the system. Invariably in such situations those trying to create the quality procedures get lost in the desire to 're invent the wheel.' As a result, personnel who, in all probability may well have been doing the job for years, find themselves becoming embroiled in what to them is needless, time consuming, and unproductive paper work and when this happens, frustration and disenchant-ment invariably follow.

So there it is. Two possibilities exist. On the one hand the creation of an evil necessity which has been mandated by the customer, the chance to win what then becomes a meaningless award, and or an opportunity to reduce the effectiveness of a company as a result of a disenchanted workforce. On the other hand the production of a systematic and formalized way of working designed to ensure that a 'quality' approach is taken in all areas of the company's operations, which in turn will create a situation where the company will produce a product or service which will satisfy the requirements of both quality and the law.

List of Cases

Adams v. Lindsell (1818) 27
Alec Norman Garages v. Phillips (1985) 141
Andrew Bros (Bournemouth) Ltd. v. Singer & Co (1934) 59
Arcos Ltd v. E A Ronaasen & Son (1933) 131

Bannerman v. White (1861) 44
Barkway v. South Wales Transport Co Ltd. (1949) 80
Barnett v. The Chelsea and Kensington Hospital Management Committee (1968) 73
Bartlett v. Sidney Marcus Ltd (1965) 133
Baxters (Butchers) v. Manley (1985) 141, 149, 159, 198
Beale v. Taylor (1967) 131
Bettini v. Gye (1876) 50
Birch v. Paramount Estates Ltd. (1956) 45
Blyth v. Birmingham Waterworks (1856) 71
Bolton v. Stone (1951) 75
Bowater v. Rowley Regis Corporation (1944) 81
Brinkibon Ltd. v. Stahag Stahl und Stahlwarenhandelsgesellschaft mbH (1983) 25
Brogden v. Metropolitan Railway Co.(1877) 54
Bunge Corporation v. Tradax Export SA (1981) 51
Butler Machine Tool Company v. Ex-cell-O Corporation (1979) 53, 186
Byrne v. Van Tienhoven (1880) 23

Carlill v. Carbolic Smoke Ball Company (1893) 16, 29
Cartlidge v. Jopling & Sons (1963) 104
Re Casey's Patents, Stewart v. Casey (1892) Casey (1892) 32

Central London Property Trust Ltd v. High Trees House Ltd (1947) 38
Chapleton v. Barry Urban District Council (1940) 57
Combe v. Combe (1951) 39

D&C Builders Ltd v. Rees (1966) 37
Dick Bentley Productions Ltd. v. Harold Smith (Motors) Ltd.(1965) 47
Dixon v. Barnett (1989) 144
Denard v. Smith and another (1990) 144, 149, 190
Dickinson v. Dodds (1876) 24
Doble v. David Greig Ltd (1972) 153
Donaghue v. Stevenson (1932) 2, 66, 91
Donnelly v. Rowlands (1970) 143

Easson v. London & North Eastern Railway Co (1944) 80
Edwards v. Skyways Ltd (1964) 44
Entores Ltd. v. Miles Far East Corporation (1955) 24
L'Estrange v. Graucob (1934) 57

Fisher v. Bell (1961) 142
Fletcher v. Bugden (1974) 141
Foakes v. Beer (1884) 36
Frost v. Aylesbury Dairy Co. (1905) 91

Grant v. Australian Knitting Mills (1936) 68, 127, 134
Godley v. Perry (1960) 135
Grainger and Son v. Gough (1896) 19
Griffiths v. Peter Conway Ltd (1939) 134

Haringey London Borough v. Piro Shoes Ltd., (1976) 142
Harris v. Nickerson (1873) 16
Hartley v. Ponsonby (1857) 34
Harvey v. Facey (1893) 15
Hollier v. Rambler Motors (1972) 62
Home Office v. Dorset Yacht Co Ltd.(1970) 69
Hongkong Fir Shipping v. Kawasaki Kisen Kaisha (1962) 50
Holwell Securities Ltd v. Hughes (1974) 28
Hughes v. Metropolitan Railway Co Ltd. (1877) 37
Hutton v. Warren (1836) 48
Hyde v. Wrench (1840) 22

Index